# Present Day Primers

## WIRELESS TELEGRAPHY

### AND

## WIRELESS TELEPHONY

## PREFACE TO SECOND EDITION.

Since the first edition of this book went to press, much development has taken place in wireless telegraphy. The range distance to which messages have been carried has increased. The number of ship and shore stations has approximately doubled. The number of wireless messages transmitted has greatly increased. A very notable service recently rendered by wireless telegraphy has drawn public attention forcibly to its value as a means of protecting lives at sea. Early on the morning of January 23rd, 1909, the east bound White Star Liner " Republic " was injured by collision with the west bound Italian Liner " Florida," about fifteen miles south of Nahtucket light ship, in a dense fog. A hurried wireless general call for assistance brought several vessels to the rescue and, in particular, the White Star Liner " Baltic," that happened to be sixty-four nautical miles west of the scene of accident. So dense was the fog that the

PREFACE

Baltic steamed for twelve hours over a zigzag
course of some two hundred nautical miles before reaching the helplessly injured and drift-
ing vessel. Even then the search would probably have been futile, if wireless messages between the ships and the shore station at Sia-
conset had not assisted the meeting. About
sixteen hundred persons were ultimately transferred to the Baltic from the Republic, which
shortly afterwards foundered in deep water.
No loss of life occurred except in the actual
collision.

In spite, however, of the above achievements
of wireless telegraphy during the past two
years, the development of wireless telephony
has been still more pronounced. During that
time, the range distance of wireless telephony
has remarkably increased; so that, although
far below that of wireless telegraphy of the
present time, it is comparable to that of wireless telegraphy ten years ago. It has, therefore, seemed desirable to add to this book several chapters on wireless telephony, while
bringing the subject of wireless telegraphy up
to date.

Wireless telephony differs from wireless
telegraphy in details rather than in fundamen-

## PREFACE

tals; but as an achievement of the human race, the transmission of the voice to great distances on the ripples of electromagnetic waves is, in one sense, a far greater extension into free space of the range of individual personality than any form of wireless telegraphy thus far attempted.

Although it has been the endeavor to present to the reader the fundamental and essential principles both of wireless telegraphy and wireless telephony, rather than any description of particular systems or inventions; yet in treating of wireless telephony, much has been taken directly from the publications of Prof. R. A. Fessenden, who has done so much to extend the knowledge and practice of wireless telephony in America.

The author also desires to express his indebtedness to the writings of Messrs. B. S. Cohen, G. M. Shepherd, and Ernst Ruhmer.

Cambridge, *Mass., February,* 1909.

# CONTENTS

# CHAPTER I

INTRODUCTORY

ELECTRICITY is *not* in its infancy, popular impression notwithstanding. Electric applications, such as the telephone, the wire telegraph, the electric lamp and the electric motor, are very familiar in modern life and have been for a number of years. Electricity has reached adolescence in these directions. But wireless telegraphy, the most recent of electric applications, is, perhaps, in its infancy. It is only about ten years old.

There is something fascinating about the way in which electricity works. It is as swift as it is stealthy. Electric impulses move over wires at enormous speeds and yet the action is invisible and inaudible, appealing to no sense directly. A telegraph wire runs overhead say from New York to Buffalo and the New York sending operator closes the circuit at his key. Almost instantly, and certainly within about one-tenth of a second, the little lever of the receiving instrument at Buffalo responds. The electric impulse reaches its destination perhaps as rapidly as we can turn our thoughts from one end of the wire to the other.

We cannot see the electric impulse rush along the wire, we cannot hear it travel. We can only picture the transfer in our imagination. We can see the wire and we know it travels along that. If we cut the wire anywhere, the electric current is stopped.

When we turn to wireless telegraphy, however, we are deprived even of the consolation of a guiding wire to aid the imagination. The phenomenon of the wire telegraph is a mystery, but a familiar one to which the wire is a clue. The new phenomenon of the wireless telegraph is a yet more elusive mystery with no clue, at first sight. Nevertheless, we shall see in the sequel that there is not so much difference between the two cases as might at first be supposed. The relations between wireless telegraphy and wire telegraphy resemble the relations between sound distributed in the open air, and sound channeled, or confined, within a speaking tube.

# CHAPTER II

SINCE wireless telegraphy is carried on by means of electro-magnetic waves, it is desirable to examine into the nature of those types of waves with which we are more familiar, before taking up the consideration of the less familiar electro-magnetic waves.

## Free Ocean Waves

A wave is a progressive disturbance, or a disturbance which moves along through some kind of medium. The most familar type of wave is the disturbance in the level of water by some displacing force, such as wind, or the splash of a falling body. Even in calm weather, we usually find a wave-motion, commonly called *swell*, on the surface of the ocean. We then find displacements of the ocean level, alternately up and down, or high and low, moving along the surface. These waves have a certain direction, say from west to east, in the horizontal plane or plane of ocean level. They also have a certain speed in this direction. On the deep ocean this speed may be, say, 12 meters per second (39.4 feet per second, or 26.8 miles per hour). Of course this does not mean that in calm weather a cork, or lifeboat, floating on the sea, is moved along by the swell at this high speed. The cork, or the boat, bobs up and down on the swell, with only a very small movement in the swell's direction. But it means that a torpedo boat would have to steam in this case at a speed of 26.8 miles per hour from west to east in order to keep abreast of any one particular roller in the swell. It is to be observed that the waves advance through and over the water, without carrying the water bodily along to any marked extent.

Another noteworthy point in the waves of ocean swell is that there is commonly a fairly definite and uniform wave-length. This wavelength is measured in the direction of wave-motion, and may be taken as the distance from crest to crest of successive waves. This wavelength in open ocean swell is often about 100 meters (328 feet or 109.3 yards). A tightly stretched string 328 feet long, held east and west, would in such a case just span the depression between successive roller crests.

The waves of the ocean manifestly contain *energy*. That is to say, work had to be done originally by the winds to produce them, and the waves are capable of doing a good deal of work before being brought to rest. Attempts have been made, in fact, to obtain work from waves at suitable points on the sea-coast by means of engines operated by the rise and fall of the waves.

## Sound-Waves in Air

A type of wave of which we are constantly receiving impressions through our ears, but which is more difficult to analyze than the ocean wave, is the sound-wave in the atmosphere. Waves of sound are invisible and hence the difficulty we experience in becoming familiar with their forms, speed and other properties. We learn that sound in air is a disturbance in its density and pressure which moves through the air at a definite speed.   If we fire a pistol in the air, the explosion in the barrel displaces the air, or compresses it, in the immediate neighborhood of the discharge.   The zone of compression moves off into surrounding air with substantially the same speed in all directions, if the air is calm, and is followed immediately by a zone of expanded air;  just as a hollow or depression follows a hump or elevation in an ocean wave.

Fig. 1 gives a diagrammatic view of a single sound-wave of compression and dilation, shortly after the wave has moved off from the explosion

FIG. 1.—Diagram Indicating a Single Sound Wave Shortly after Leaving Its Source.

or origin of the disturbance, o. The wave is hemispherical in form, and the diameter of the hemisphere is *a a*, at the instant considered. The external portion of the hemispherical wave shell contains compressed air, indicated by the concentric semi-circles. The internal dotted portion, immediately following, contains expanded air. After a brief interval of time the wave will have expanded to the contour indicated in Fig. 2, where the diameter of the hemis-

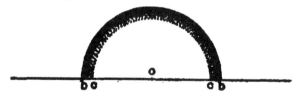

FIG. 2.—Diagram Indicating a Single Sound Wave Which Has Separated Itself from Its Origin by Five Wave-Lengths.

pherical shell is *b o b*. The length of the wave is the distance *b c*, measured radially from the front to the back of the wave, or from the out-

side to the inside of the hemispherical shell.   If
this wave-length be, say 100 meters (328 feet or
109.3 yards), then the distance *o c*, by which
the wave has already removed itself from the
origin, is 500 meters (1640 feet or 546.7 yards).
Another brief interval of time would bring about
the condition indicated in Fig. 3, where the wave

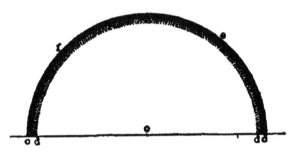

FIG. 3.—Diagram Indicating a Single Sound-Wave
Which Has Separated Itself from Its Origin by
Eleven Wave-Lengths.

has expanded hemispherically to the diameter
*c o c*.   The length of the wave, *c d* remains as it
was in the earlier stages, assumed as 100 meters
(328 feet or 109.3 yards); but the distance *o d*
which the wave has now placed between itself
and the origin is 1100 meters (3609 feet or 1203
yards).   Observers in balloons floating in the
air at points such as *e* or *f* on the wave front,
would hear the sound of the explosion simultaneously with observers on the ground at
points *c c*.

## Speed of Sound-Waves

The speed with which the sound-waves moves radially outward in all directions is approximately 333 meters per second (1090 feet per second, or 746 miles per hour), depending slightly upon the temperature and humidity of the air. We cannot see the hemispherical shell of disturbed air expanding; but we can picture the process to the mind's eye. In an hour, the expansion would carry the radius of the hemisphere to a distance of 1200 kilometers (746 miles). But the density of the atmosphere would be infinitesimally small at an elevation amounting to this distance, and sound being a disturbance of air cannot travel where the air ceases to exist. Consequently, the hemispherical form of the wave must disappear at great distances, for lack of air above the origin.

## Dilution of Intensity in Sound-Waves

The energy residing in the sound-wave would be the same in the successive states of Figs. 1, 2, and 3, if there were no expenditure of energy in friction during the motion. That is to say, we may suppose that a certain part of the energy in the explosion at $o$ was stowed away in the wave of compressed and dilated air. But the space occupied by the wave in the stage of Fig. 2 with radius *o a*, 150 meters (164 yards), would be 13.6 millions of cubic meters; in the stage of Fig. 2, 382 millions of cubic meters, and in the stage of Fig. 3, 1660 millions of cubic meters (corresponding to 17.8, 500, and 2160 millions of cubic yards respectively). It is evident, therefore, that the energy in the wave is constantly spread out into more space, or diluted, as the wave expands; so that the energy in a given volume, such as 1 cubic meter, is constantly diminishing. This is another way of saying that the *intensity* of the wave diminishes as time goes on, and the radius of the wave increases. The loudness of the explosion as noted by an observer at *a* in Fig. 1 would be considerably greater than that noted by an observer at *b* in Fig. 2; or again, than that noted by an observer at *c* in Fig. 3.

If we were to place a sufficiently sensitive recording barometer anywhere in the neighborhood of the explosion, and carefully observe the barometer record as the noise of the explosion occurred, we should expect to find that the barometer would record, just before the explosion, a horizontal straight line *a b*, Fig. 4, corresponding to the reading of the barometer at that time, say 760 millimeters (29.92 inches) of mercury. When the sound of the explosion arrived, the barometer would rise very slightly to *b c*, then fall to *e*, and then return to the normal straight line *f g*. The elevation from *b* to *c* would mark the degree of compression in the first half of the wave, and the depression from *d* to *e* would similarly mark the following dilatation. The crest height *k c*, or *k₁ e*, would measure the *amplitude* of the disturbance or amplitude of the wave. If the barometer were located close to

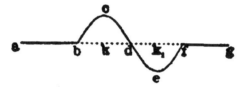

FIG. 4.—Ideal Barometric Record at the Time of Passage of the Noise of an Explosion.

the origin of the explosion, as at *a* in Fig. 1, the amplitude of the pressure disturbance record, and the amplitude of the recorded sound, would be relatively large. If, on the other hand, the barometer were placed further from the origin, as at *c* in Fig. 3, the amplitude both of the recorded disturbance and of the sound-wave at the barometer would be smaller.

Ordinary sound-waves possess so little energy, or have so small an amplitude, that recording barometers show no trace of them. Expressing the same thing in another way, the impression-producing mechanism of the ear is far more sensitive to the disturbances of pressure in sound-waves than the ordinary barometer.

Very powerful explosions are capable of producing sound-waves of sufficient intensity to be observed at great distances. The great explosion of the volcano Krakatoa, near the Sunda Straits, in the year 1883, is stated to have been heard at distances greater than 3200 kilometers (2000 miles). The outgoing wave affected barometers all over the world, and left traces on recording barometers. This wave is stated to have traveled at a speed of 1130 kilometers per hour (700 miles per hour), to have swelled at the antipodes to Krakatoa, in 18 hours, and to have spread out again over the globe. It was not finally lost sight of until it had passed around the globe several times. When the first outgoing wave passed over Singapore, a port distant about 830 kilometers (516 miles) from the origin of the disturbance, the gas-holder of the town is stated to have leaped into the air several feet up and down.

### Sound-Wave Trains

If instead of producing a solitary explosion at the origin, and a corresponding solitary wave moving off radially therefrom, we produce a

rapidly and rhythmically repeated disturbance;
as in blowing air through an organ-pipe, or
forcing air through a syren-wheel, a succession
of waves is produced, and the sensory effect
produced on the ear is that of a tone or musi-

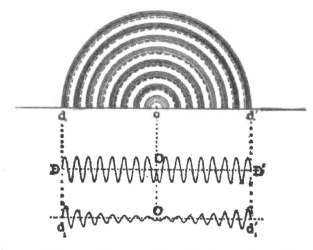

Fig. 5.—Diagram Representing the Succession of Sound-
Waves Emitted from the Origin *O* When a Simple
Musical Note Is There Produced.

cal note. A succession of outgoing waves is
diagrammatically represented in Fig. 5. The
shaded areas there correspond to zones of compressed air, and the intermediate unshaded
areas to zones of dilated air. Eight complete
waves are indicated. If the note sounded be
the deepest E of the double-bass viol, making
40 complete vibrations a second, then the length

of each wave will be the fortieth part of 333 meters, the distance which sound travels in a second, because 40 complete waves will occupy the space covered by advancing sound in one second. Each wave will, therefore, be 8.33 meters (27.34 feet or 9.11 yards) in length, and the length *o d* of eight wave-lengths will have been covered in one-fifth of a second from the commencement of the sound.

The curve at D O or O D' gives the trace-record that we should expect a recording barometer would give at *d* after all of the eight waves passed by, on the supposition that this pure, musical note was sustained uniformly for eight complete cycles of the disturbance. If, however, the string of the double-bass, instead of being excited by the bow, were plucked by the finger in such a manner that the vibrations died away, then the record of the supposed sensitive barometer at d might indicate the curve d, o of decaying amplitudes.

# CHAPTER III

HAVING paved the way for a consideration of electromagnetic waves by a few outlines of sound waves in air, we may now fitly turn attention to magnetism and electricity.

## Wind and Its Energy

Everyone is familiar with the fact that wind is an active or disturbed state of the atmosphere, a movement of the air. We ordinarily understand wind to be a uniform movement of the air in any one given direction, and we ordinarily understand by *eddies* or *gusts*, twisting or vortical movements of the air, but, in general, wind may include both linear movement and vortical movement, since one cannot occur in the atmosphere without involving the other. The material for the creation of a wind is always present, for this material is the air itself. We only need to energise the air in a particular way, to make it move forward. Energy must be expended in producing a wind, and energy resides in the wind.

If we employ a hand-fan to produce a local breeze, we must expend muscular energy, or do work on the fan, to force the air into motion, and the air once set in motion contains energy or can do work by moving, for example, light obstacles in its path. Consequently, we may say that wind is air, plus energy given to it in a particular way.

Air is a material fluid. It forms an ocean on the surface of this earth, and we live at or near the bottom of this air-ocean. Air gravitates, or pulls upon the mass of the earth. Each individual atom of air gravitates, and the sum total of all the individual pulls exerted on the earth amounts to a pressure of about 1 kilogramme per square centimeter of surface (14.25 lbs. per sq. inch).

### The Invisible Ether

It is generally believed that all space, including the interior of solid bodies, is permeated by an immaterial fluid called the universal ether. The ether is just as invisible as air. Whether it consist of matter or not, it is immaterial in the sense that it apparently does not gravitate. It does not directly appeal to any sense, but it is much easier to assume its presence everywhere than to deny its existence. If we take a vacuum-tube, *i.e.*, a sealed glass tube from the interior of which the air has been almost entirely removed, it can be shown experimentally that sound cannot move across the interior of the tube, but light passes across it, and so do radiant heat and gravitational force. We cannot believe that these activities are transmitted through absolutely empty space. Something must transmit them, for they are transmitted at definite speeds. This something is named the ether. Beyond its powers of transmitting energy, hardly anything is yet known about the ether. Its structure, and the manner in which it permeates space, are still unsolved riddles.

## *Nature of Electricity and Magnetism*

As soon as the ether is postulated to be a universal fluid or medium in which all matter swims, so to speak, many things may be accounted for which otherwise we could not even attempt to explain. Electricity and magnetism, for example, may be accounted for in a general way. Just as wind is, we have seen, a particular energized condition of the circumambient air, so both electricity and magnetism are particular energized conditions of the universal ether, which underlies the air and everything else. It is not so easy, however, to define the nature and relations of these particular energized conditions. We cannot at present say, for example, that electricity is the same kind of motion of the ether that wind is of the air. If we do not yet apprehend the nature of the ether itself, how shall the task be undertaken of defining its energized conditions? The energized conditions might be statical, and involve no motion of ether, like the energy of a stationary coiled-up spring; or they might be dynamical, and involve modes of motion of the ether. In any event, it seems clear from the known laws of electromagnetism that there is a definite mutual relation between the two energized conditions of ether, electricity and magnetism, such that as soon as either is defined the other also is immediately determined. In mathematical language, one is the "curl" of the other. If, for instance, electricity should be a definite kind of tension or static stress longwise, then magnetism would be a definite kind of twist or crosswise static stress, and reciprocally. Or, if electricity should be a simple, straightforward motion, or streaming, of the ether, then magnetism would be eddy motion or vortical rotational motion, or spin, of the ether, and reciprocally. It is surprising how much is known concerning the laws of action and behavior of electricity and magnetism, considering the little that is known of their absolute fundamental nature. We can control electricity and magnetism remarkably well, considering that we do so from beyond a hitherto impenetrable veil that does not admit of perceiving the things directed.

It is evident that whatever may be the precise nature of electricity and magnetism, the widely admitted postulate of the universal ether requires that the material for either or both is omnipresent. Just as the material for wind is always present in the circumambient air, and all we need is the application of energy to the air in a particular way in order to produce a wind, so the material for electricity or magnetism is universally present, and all we need is the application of energy to the ether in particular ways. Consequently, electricity and magnetism may be regarded as the ether, plus particular forms of energy.

### Magnetic Flux and Its Properties

If we consider an ordinary, permanent horseshoe magnet, such as is indicated in Fig. 6, we find that all around it, and particularly between its poles N and S, there is a certain invisible activity which possesses both direction and intensity. In the illustration, the direction is roughly indicated by the broken lines with arrow-heads, and the intensity by the relative crowding or condensation of these lines. The direction of this magnetic activity in the air between the poles is seen to be from the north pole N to the south pole S. This is strictly speaking a pure convention. It might have been originally agreed to draw all the arrows in the opposite direction. All that is certain is that there is a definite polarity about the system, and that the actions pertaining to the north pole are distinctly inverse to the actions pertaining to the south pole, magneticians all agreeing upon the direction shown. The north pole is the pole which, if the magnet were freely suspended, would seek for, or point approximately toward, the north geographic pole of the earth, or the earth's magnetic pole near the Greenland end of the earth's axis. That is, the N end is the north-seeking end.

Fig. 6.—Diagram of Magnetic Flux in the Space Between the Poles of a Permanent Magnet.

As roughly indicated in Fig. 6, the magnetic activity, or *magnetic flux* as it is called, is densest or most intense, in the air between the opposing

pole-tips, or where the air-space separating the
poles is shortest. As we leave this region, the
magnetic flux becomes thinner, or weaker. A
peculiarity about this flux is that it always re-
turns back upon itself in closed loops or chains.
In other words, magnetic flux is always continuous and re-entrant. At first sight it appears
to be discontinuous, because it seems to commence at one pole and end at the other. But it
can be shown experimentally that the flux continues through the interior substance of the steel
magnet, and each loop, such as N A S, completes
a circuit B C D E F within the substance of the
magnet.

### Provisional Hypothesis as to Nature of Magnetic Flux

Although the real nature of this magnetic flux
is not yet known, yet it may help us to follow the
actions of electromagnetic waves later on, if we
assume, for the purposes of description, that
magnetic flux is a streaming motion of the ether.
On this assumption, a permanent magnet is a
force pump which draws the ether in at the
south pole, through the substance of the steel
in the interior, and forces it out at the north pole.
With no friction, this streaming would not necessarily absorb energy, and we know that perma-

nent magnets may be designed to retain their magnetism without sensible diminution for an indefinitely long time.

## *Energy of Magnetic Flux*

Although magnetic flux does not need energy to be expended in order to keep it going, yet energy has to be expended to create it. That is, magnetic flux contains energy, or has energy always associated with it. As long as the magneitc flux persists, the energy resides quiescent with it. When the flux disappears, its energy disappears also. Consequently work must be done to create magnetic flux, and magnetic flux is able to do work or give up its energy when it disappears.

Between the opposed pole-tips N S, we may consider, on the above hypothesis, that the stream of ether is densest, and receding from this region the stream gets weaker. The streaming is steady both at any particular point for all considered time, and for all points at any one time. The magnet ether-pump is steady. The pumping action is due to activities in the molecules of the steel. Each molecule of iron is supposed to be a little individual ether pump, by virtue of internal activities as yet only dimly guessed at. When the horseshoe is magnetized, all of the molecules are caused to align themselves in parallel directions, or to face the same way, whereby they all pump the ether in the same general direction. Within the iron molecules, the pumping activities are believed to be electric; but into these we need not enter. The point here to be observed is that in the air-space outside of the magnet, the steady magnetic flux produces no electric action. In this air-space we find magnetism but not electricity. If, however, we move the magnet, and with the magnet the system of magnetic flux pertaining thereto, there will be electric action produced where the magnetic flux lines are carried through space. If, for example, the magnet be lifted bodily toward the observer without twisting, feeble electric forces will be brought into play in directions lying within the plane of the horseshoe. In the region between the poles these electric forces will be directed, during the motion, in the direction from A to D. The intensity of these electric forces will be proportional to the speed with which the magnetic flux moves sidewise. If the magnetic flux moves longwise, or parallel to itself, there is no electric action set up, but if the magnetic flux moves sidewise, or crabwise, there is electric action set up. It is on this action that all dynamos depend; namely, upon relative sideways motion between magnetic flux and an electric conductor to pick up and utilize the induced electric force. Steady and stationary magnetic flux is thus unaccompanied by electric action, but unsteady, varying, or sidewise-moving magnetic flux sets up electric action.

### Electric Flux and Its Properties

Turning now to electricity, Fig. 7 represents a vertical metallic rod, and terminal balls, sup-

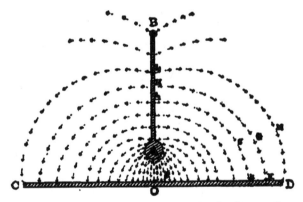

Fig. 7.—Diagram of Electric Flux in the Space Between an Electrified Rod and Disk.

ported by an insulating holder not shown, in air above the center O, of a horizontal, insulated metallic disk C O D. This insulated pair of conductors may be electrically charged either by a frictional electric machine, an influence machine, a spark coil, or a voltaic battery. That is, the charge may be communicated from any suitable electric source. The charge will be retained, because the rod is insulated from the disk, and if the insulation could be made perfect, the charge would be retained indefinitely. Let us suppose that the rod is positively charged and the disk negatively.

In the air-space between the electrified rod and disk there is an invisible influence which possesses both direction and intensity. It is called electric flux. This electric flux is diagrammatically represented in Fig. 7 by the little arrows which proceed, by convention, from the positive charge to the negative charge. The arrows are drawn on little lines of points, instead of little broken lines as in Fig. 6, in order to distinguish them from lines of magnetic flux. Between A and O, where the air-space is shortest, the electric flux is most densely crowded, or its intensity is greatest. As the separating air-space increases, the flux density weakens.

### Energy of Electric Flux

Energy always resides in the electric flux, so that each and every part of the region permeated by the electric flux represented in the illustration contains energy. The energy is not uniformly distributed. It is greater per unit volume at a

point like F than at a point like G.   It is stowed
away in proportion to the square of the electric
flux density, so that in two regions one of which
has double the flux density of the other, there
will be four times more energy per cubic centimeter, or cubic inch of space, in the former than
in the latter.   As long at the electric flux persists, this energy resides therein or accompanies
it, and when the flux disappears the energy has
disappeared.   This energy is communicated to
the ether in the insulating air between rod and
disk at the time of their charge.

### Provisional Hypothesis as to the Nature of Electric Flux

In conformity with the provisional hypothesis
already adopted for magnetic flux, stationary
electric flux may be assumed to be an elastic
twist or stress in the ether; so that the whole
system of ether tends to revolve clockwise about
the rod A B as axis, when looking down on the
disk from above.   The screw or twist will have
maximum intensity along the central line O A,
and is resisted by the elastic rigidity of the ether.
The elastic energy of the twist is the total energy
of the electric flux as summed up throughout
the entire *electric field,* or permeated insulating

region. The amount of electric energy that air can be made to hold without breaking electrically, or disrupting into a spark discharge, depends upon the atmospheric pressure and upon the shape of the opposed electrified surfaces. At ordinary atmospheric pressures, and parallel opposed surfaces, the most favorable form, the energy that air can hold is limited to about 480 ergs per *c. c.*, or 1 foot-pound per cubic foot; *i.e.*, the work done by lifting one pound one foot high, to the cubic foot of air space under powerful electric stress.

Electric flux at rest differs from magnetic flux at rest in the fact that the former is discontinuous while the latter is continuous. The magnetic flux, as we have seen, always forms closed loops or chains in space. Steady electric flux, on the contrary, always starts from a positive charge and ends on a negative charge. In the case of opposed conductors, the charges always reside on their surfaces, and thus the electric flux always starts on the surface of the positive conductor and ends on the surface of the negative conductor.

### Tensions in Electric and Magnetic Fluxes

The electric flux, like magnetic flux, always possesses the property of pulling along its own

direction at the same time that it pushes sideways. The curved arrow lines of Fig. 7 merely
indicate the direction and the relative crowding
of the electric flux from point to point of this
particular electrified system; but if we suppose
that each of these lines is a little elastic thread,
exerting a certain mechanical tension, and if we
also suppose that each such elastic thread exerts
a repulsion sideways against its neighbors, or
tries to secure all the elbow-room it can, we get
an idea of the static forces which reside in such
a stationary electric flux. Thus, the line H G K,
in addition to its own tension, pushes sideways
against the adjacent lines h g k and L M D.
The resultant effect is a tension, or attractive
pull, between the rod and the disk, or the familiar
attractive force between oppositely electrified
bodies.

As long as the electric flux remains steady and
stationary, no magnetism, or magnetic force is
produced. There will be a tendency to move
any electrified object, such as a pith-ball, along
the electric flux in Fig. 7, but there will be no
tendency to affect the direction of a magnetic
compass-needle. If, however, the electrified
system be moved bodily sidewise, without losing
the charge, feeble magnetic forces will be developed in directions at right angles both to the

moving electric flux and to the direction of motion.

Just as sidewise-moving magnetic flux generates electric flux, so sidewise-moving electric flux generates magnetic flux.

# CHAPTER IV

## ELECTROMAGNETIC WAVES GUIDED OVER CONDUCTORS

### *Automatic Movement of Electric Flux over Conducting Surfaces*

IN order to bring electric flux into movement, it is not necessary to move a charged system of conductors, the flux will set itself in motion if an opportunity is offered to let it slide upon a pair of conductors.

### *Electric Current Over a Pair of Wires*

If we bring a long pair of parallel insulated metallic wires M N, P Q, Fig. 8, simultaneously into contact, one with the rod and the other with the disk, as indicated in the figure, the electric flux immediately takes advantage of the extension of the system thus offered and glides away, guided by the wires. It may be considered that the sidewise repulsion of the flux tends to make it spread its boundaries in this way, whenever possible. The electric charge moves out along the wires hand in hand with the electric flux, the positive charge spreading along the upper wire

M N, in Fig. 8, and the negative charge along the lower wire P Q. The electric flux runs along with these charges, always bridging over between the positive and negative charges. The phenomenon of the movement of two parallel moving

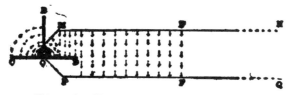

Fig. 8.—Electric Flux Wave Moving Over Pair of Parallel Insulated Wires.

charges with the moving electric flux between them and linking them, constitutes an *electric current*, or electric discharge.

The effect of bringing the two parallel wires into contact with the charged electric system of Fig. 7, is, therefore, to let the charge escape over the wires, and to set up thereby an electric current over the wires. The current rush takes place in the form of a wave. Electric flux and its associated energy move off the disk into the insulating air-space between the wires.

## Creation of Magnetic Flux by Moving Electric Flux

We have already seen that sideways-moving electric flux generates magnetic force and flux.

As soon, therefore, as the electric flux begins to move, half of the electric flux energy disappears and is replaced in the form of magnetic energy. Instead of having stationary electric flux in the confined insulated system of Fig. 7, we have moving electric flux, and magnetic flux associated therewith, or advancing with it.  Consequently, although we can have either stationary electric flux alone, or stationary magnetic flux alone, we cannot preserve them independently when they move freely in an insulator.  Any wave of electric disturbance is an *electromagnetic wave*, because in it electric and magnetic fluxes are tied up together.

## *Electromagnetic Wave Guided by a Pair of Parallel Wires*

The distributions of the electric and magnetic fluxes in the advancing wave of Fig. 7 are illustrated in Fig. 9, where M P are the sections of the two parallel wires in a plane at right angles to their length.  The electric flux-paths are indicated, as in previous instances, by lines of points with arrow heads; while the magnetic flux-paths are indicated by broken lines with arrow heads. The wave is supposed to be receding from the observer, and the upper wire M is carrying the positive charge, while P, the lower wire, carries the negative charge. The electric flux, there-fore, emerges from the surface of the wire M and terminates upon the surface of the wire P. If the metal of which the wires are composed be supposed to conduct perfectly, the electric flux will skim over the sur-faces of the wires and not penetrate into their mass. The more im-perfect the conductivity of the wires, the more deeply the moving elec-tric flux will penetrate into them.

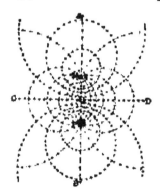

FIG. 9.—Electric and Mag-netic Fluxes in Electro-magnetic Wave Guided by Two Parallel Wires. Wave Receding from Observer.

The magnetic flux at the center O of the loop has the direction D O C, perpendicular to the loop. At all other points the magnetic flux-paths are circular in this plane of cross-section, or cylindrical with regard to a length of the wires. Both the electric and the magnetic flux-paths are systems of circles, and it is to be noticed that at every point they intersect each other perpendicularly. That is, any one circle crosses all the intersecting circles at right angles.

It is also to be observed that where the electric

flux runs densest, so does the magnetic flux. The densest electric and magnetic fluxes are found close to either wire. Both the fluxes get weaker as we recede from the wires. In fact the intensity of the electric flux in any single pure electromagnetic wave is always and everywhere numerically equal to the intensity of the magnetic flux at the same point and time, when each is measured in its appropriate units. At an indefinitely great distance from the loop of active wires the density of the fluxes is nil.

### Speed of Electromagnetic Waves

The speed of sound waves in air we have seen to be in the neighborhood of 333 meters per second, (1090 feet per second or 746 miles per hour). But the speed of a free electric wave in air is enormously greater, being approximately 300,000 kilometers per second (186,400 miles per second), or $7\frac{1}{2}$ times around the world in a second. This is also the speed at which light travels in air. That is to say, no difference has yet been determined between the speed of electromagnetic waves in air and the speed of light.

Energy is carried in the advancing electromagnetic wave indicated in Figs. 8 and 9. The energy is the energy residing in all the electric flux that moves on, plus the equal amount of energy in all the associated and interlinked magnetic flux.   This energy is carried away from the original stock of electric energy in the air-space of the electrified system in Figs. 6 and 7. The energy was originally bound up in the stationary electric flux.   The advancing electric and magnetic fluxes in the wave robbed the charged system of flux and of energy and transported that energy whithersoever they went.

Summing up the conditions which we have noted in the guided electromagnetic wave of Figs. 8 and 9 we may state them as follows:—

An electric or magnetic disturbance associated with a pair of insulated aerial wires propagates itself along the wires at the speed of light.   The wave consists of electric and magnetic fluxes, which are always perpendicular to each other and to the direction in which the wave is moving. If the two wires are parallel, the fluxes are distributed cylindrically; i.e., circularly in any plane perpendicular to the wires.   The energy in each flux is the same, and the intensity of the two fluxes is the same at every point.   The energy per unit volume varies as the square of the intensity of the moving fluxes.   The advancing wave conveys this energy with it.   On the surfaces of the wires are opposite electric charges, moving with the flux, and supporting the ends of the electric flux.   The entire series of associated phenomena is an electric current.

Guided electromagnetic waves properly belong to the domain of ordinary wire telegraphy, or to the transmission of electric power by wires.   As such, they lie outside of the province of this en-quiry.   It may suffice to observe that the steady electric current found in any electric circuit oper-ated by a dynamo, or a voltaic battery, is merely the sum of what is usually a large number of superimposed electromagnetic waves of the type above considered.   These waves are kept stream-ing out of the dynamo, and are also reflected back from the distant end, or other parts, of the circuit; so that after a brief interval of time we have a complex aggregate of waves present.   We may now proceed to the study of semi-guided electromagnetic waves, or waves in which the electric flux is held at one end only on an insu-lated artificial conductor, or is guided by but a single wire.

# CHAPTER V.

## Electromagnetic Wave Guided by a Single Wire and the Ground

IF we place the disk C O D of Fig. 7 upon the level surface of the ground, taking pains to secure good electrical conductivity in the adjacent soil, and charge the vertical rod A B, while supporting the same insulated above the center O, there will be but little change effected in the charge or distribution of the electric flux by reason of the grounding of the disk. From an electrical point of view, we shall merely have indefinitely extended the area of the conducting disk. If we now bring a single very long insulated horizontal wire M N, like an ordinary telegraph wire into contact with the charged rod, as indicated in Fig. 10, the electric charge and electric flux will immediately rush out at light-speed over this wire in an electromagnetic wave. The electric flux will be guided by the wire M N on its

36

positive ends; but the negative ends will be un-
constrained, or left loose to themselves.   In this
sense, the wave is only singly guided.   If the
surface of the earth G G be assumed to conduct
perfectly, the electric flux will skim over this surface, and a negative charge will also distribute

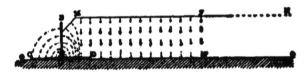

FIG. 10.—Single-Wire Guided Electromagnetic Wave.

itself over the same, advancing with the electric
flux.

The electric flux under these circumstances
will spread out over the surface of the ground
G G in such a distribution as would be effected
if the ground were removed and in its place a
second wire were run parallel to M N and as far
beneath the level surface G F G as the first wire
M N is above it.   The distribution is indicated
in Fig. 11, where M is the section of the wire in a
plane at right angles to its length, and G G is the
conducting surface of the ground.   The wave in
this case is supposed to be advancing towards the
observer.   The lines of points show the paths of
electric flux issuing from the positive charge
moving along the wire M.   They terminate at a

negative charge distributed over G G, and moving over the same with like rapidity. N is the position of the "image" wire, which, in the absence of the ground G G, would be able, as in Fig. 9, to produce the same distribution of fluxes above the level G O G, as is developed with the

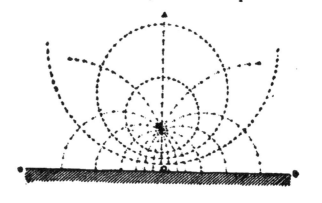

FIG. 11.—Electric and Magnetic Fluxes in Electromagnetic Wave Guided by a Single Wire Over a Conducting Ground Surface.   Wave Advancing Toward Observer.

conducting ground.  If the ground conducted perfectly, the electric flux would not penetrate below the surface; but would slide frictionless over the same.  In practice, the conductivity of the ground is never perfect and the fluxes penetrate beneath the surface to a greater or less extent, with a corresponding expenditure of energy in the soil.  Nevertheless, the conditions are ordinarily regarded as a slight deviation from the condition of perfect earth conduction as indicated in the Figure.

The magnetic flux is established in cylindrical distribution, as shown, by the motion of the electric flux at the light-speed.  The two fluxes have equal densities and energies at any given point, and between them they transport to a distance, along the wire, the energy originally bound up in the stationary electric flux of the charged system of Fig. 7.

The process thus initiated pertains to single-wire telegraphy, the usual type of wire telegraphy. The currents employed in telegraphy consist of such electromagnetic waves, either singly, or in superposition by confluence.

# CHAPTER VI

## Radiation of Electromagnetic Waves by an Electric Disturbance or Explosion

THE electromagnetic waves considered in the last two chapters were set in motion by bringing a wire, or a pair of wires, into electric connection with the charged electric system, and allowing the electric flux to overflow from that system on to and along the wires. But electromagnetic waves may also be set in motion by sudden disturbances of an electric charge. In such cases the emitted waves are likely to be radiated in all directions in a manner resembling the expansion of a sound wave in air as outlined in Chapter II.

Let us suppose the rod and disk system to be charged, as indicated in Fig. 7, after the disk has been placed horizontally upon the surface of good conducting ground. Instead of allowing a wire to approach the rod and discharge it, let the system be discharged by a spark between A and O (Fig. 7). Let us assume that the discharge is completed by a single spark of extremely short duration; so that the entire system of electric flux collapses precipitately. The flux near the axis A O is the first to disappear into the spark, then the longer and outlying flux. Last of all, the longest and furthest reaching flux issuing from B, will run down the rod and vanish at the gap A O.

If the discharge be delayed, or the charge allowed to dribble slowly across the gap A O, as, for example, by the action of rough and oxydised opposing surfaces, the collapse of the entire umbrella-shaped electric flux system of Fig. 7 may take place without any appreciable electromagnetic disturbance. The energy stored away in the flux will be expended in heating the path of discharge, and, when the process is complete, the flux having disappeared, the discharge stops, and there is no aftermath.

If, however, the collapse of the umbrella flux system in Fig. 7 is sudden, the rapid descent of the flux down the rod and into the gap A O sets up magnetic forces and magnetic flux. We have already seen that when electric flux moves, it establishes magnetic flux in a direction across itself, and also across the direction of motion; also that energy is imparted to the magnetic flux at the expense of the electric flux energy. If we

consider, for example, the particular flux path
L M D in Fig. 7, it is evident that during the
brief interval of collapse it tends to run at light-
speed into the successive positions H G K, h g k,
and so on to A O, when it disappears into the
spark discharge.   But the movement of this flux
element will set up magnetic flux directed in
planes parallel to the disk, and pointing towards
the observer.   On the other side of the rod A B,
the similar inrush of electric flux will beget a
magnetic flux in horizontal planes, *i.e.*, planes
parallel to the disk, but directed away from the
observer.   Considering all the actions that occur
simultaneously, it will be evident that a concentric ring system of magnetic flux will be set up,
as in Fig. 12, around the rod A B as axis, each
ring being in a horizontal plane.   Part of the
energy of the original electric flux of Fig. 7 is
delivered to this ring distribution of magnetic
flux; so that when the discharge is sudden, a
lesser total energy tumbles into the spark than
when the discharge is slow.

Fig. 12 indicates, in plan view, the ring distribution of magnetic flux accompanying the
collapse of the umbrella electric distribution of
Fig. 7, on the passage of a spark at O A.   The
eye of the observer is supposed to be situated
immediately over the disk C D and rod B.   The

magnetic flux streams are all directed clockwise, and they lie in various horizontal planes.

According to our provisional theory, we found in Chapter III, at page 25, that the original charge of the rod-and-disk system of Fig. 7 gave

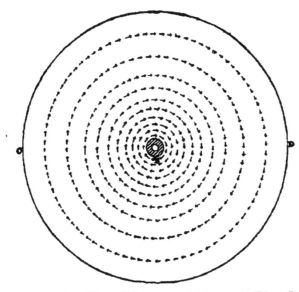

FIG. 12.—Plan View Diagram of Magnetic Flux Distribution Accompanying Collapse of Electric Flux
Around Charged Rod.

a right-handed or clockwise screw-twist to the surrounding ether, as viewed by an observer looking down on the disk. The ether may be supposed to have taken a right-handed, or clock-wise, elastic set or strain, under the stress of the electric flux, which stress is supposed to be resisted by the elastic rigidity of the ether. The stress is a maximum at the axis O A. If the ether gives way, by the disruption of the air particles at this point, the rigidity fails to oppose the stress, and the ether flows clockwise bodily in the magnetic stream lines of Fig. 12. It comes to the same thing, therefore, whether we view, in imagination, the collapsing electric flux during the sudden discharge of the system, and watch the magnetic flux rings spring into exist-ence with the downward electric motion, or whether we view in imagination the ether give way under the screw twist of the original charge, and watch the flow of the ether in obedience to that twist when the spark occurs at the axis.

### Electromagnetic Wave Generated by Sudden Collapse of Electric Flux Distribution

The ring magnetic flux, as in Fig. 12, accom-panies the collapsing electric flux down the rod; it also sets up, by reaction, an external wave of upward and outwardly rising magnetic ring flux in the counter-clockwise direction. This rising magnetic flux sets up in its turn electric flux across itself. The direction of this rising shell of electric flux is indicated in Fig. 13 at A C and A D. It is directed from the rod to the disk, as in the original charge distribution of Fig. 7.

This hemispherical shell of downward electric flux, and ring magnetic flux, expands radially outwards in all directions at the light-speed.

The collapsing ring magnetic flux of Fig. 12, when it reaches the disk, is reflected back and up

FIG. 13.—Electric Flux Induced by the Ring Distribution of Magnetic Flux in Fig. 12.

the rod, still clockwise in direction, but moving upwards immediately behind the shell C B D. It sets up an electric flux in the directions indicated at A, Fig. 13. The two concentric hemispheres of electric and magnetic fluxes detach themselves from the rod in the manner diagrammatically indicated in Fig. 14. In the external hemispherical shell w P w, the electric flux is downwards and the magnetic flux lies in counter-clockwise rings centered on the polar axis B P. In the internal hemispherical shell x Q x, the directions are reversed, the electric flux being upwards, and the magnetic clockwise.

In a very brief interval of time after the discharge of the system by the passage of the spark, we have complete disappearance of electromagnetic charges, fluxes and energy in the rod-

.md-disk system, while a hemispherical electromagnetic wave moves off radially with the light-
speed, the radius of the hemisphere being theoretically 300,000 kilometers (186,400 miles) after
one second of time. The thickness of the

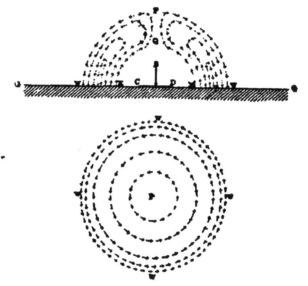

Fig. 14.—Vertical Cross-Section and Plan of Single
Expanding Electromagnetic Wave.

double-layered hemispherical shell remains constant, but since the energy in the shell also remains constant, in the absence of absorption, the
density of the fluxes and their energy per unit of
volume rapidly diminish. In other words the
energy per cubic meter of space in the wave
rapidly diminishes.

The feet of the electric flux lines skim over the surface of the ground, assumed to be perfectly conducting. At the external edge w w w w, a negative ring charge runs out radially over the ground surface at the light-speed. At x x, a similar positive ring charge runs out radially at the same speed. This succession of running electric charges, linked together by loops of electric flux, constitutes a single *cycle of alternating current* flowing along the ground.

From its external aspect, the expanding hemispherical electromagnetic wave has electric flux distributed along meridians of longitude, symmetrically disposed with respect to the polar axis B Q P. The magnetic flux is distributed in circles of latitude, the smallest circles being near the pole, and the greatest near the equator or ground surface.

### Resemblances Between Solitary Explosion Waves of Sound and Electromagnetism

The solitary hemispherical electromagnetic wave of Fig. 14 bears some resemblance to the solitary hemispherical sound wave in air of Fig. 3. Each consists of a double layer, the disturbance in the external layer being of the opposite sign to that in the internal layer. On the other hand, there are notable differences. There is an

enormous difference in speed (nearly a million
to one). The electric wave has a polar node at
P Q and the sound wave has none. The electric
wave is propagated in the ether, the sound wave
in a gaseous substance.

If the earth's surface is not perfectly conducting, and in practice it is far from being perfectly
conducting, the electric flux will penetrate to
some extent into the soil, carrying also magnetic
flux with it. The fluxes which sink in this way
expend their energy in warming the soil very
slightly and the hemispherical wave is thereby
drained of a part of its energy, or is subjected to
frictional losses in running along the ground.
The energy per cubic meter of wave shell will
thus diminish more rapidly than would be accounted for by the mere increase in bulk of the
expanding wave shell.

### Electromagnetic Wave-trains

The discharge of a rod-and-disk electrified
system does not ordinarily give rise to but a
single electromagnetic wave, such as is depicted
in Fig. 14. On the contrary, the discharge generally gives rise to a series, or train, of successive
waves of diminishing amplitude, each feebler
than its predecessor. The rate of diminution
depends upon the amount of heat expended in

the discharge spark, and to a lesser extent upon other structural details, but the amplitude of oscillation usually falls to one half, or loses fifty per cent., in about two complete swings, or after two successive waves have been thrown off. If the spark remained uniform, the amplitude would in such a case fall to one half again in two more swings, or to one quarter of the original amplitude in a total of four swings, or to one eighth in six swings and so on. Consequently, the amplitude of the successive waves emitted by a simple rod oscillator soon dwindles into insignificance.

### Analysis of Oscillatory Current on Rod and the Generation of Waves

It may be of interest to consider in some detail the process of emitting a train of hemispherical electromagnetic waves from a rod-and-disk system laid on a perfectly conducting ground and suddenly set into electric oscillation by a spark discharge. Such a system may be briefly described as a *simple vertical oscillator*. Reference is made to Fig. 15, where the vertical rod is represented in nine successive stages of the process of manufacturing and shipping half of an electromagnetic wave, and two such diagrams would be

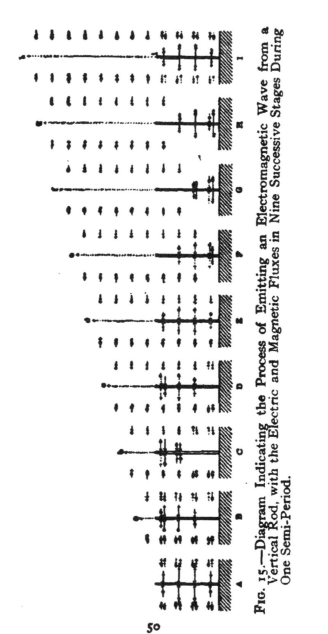

Fig. 15.—Diagram Indicating the Process of Emitting an Electromagnetic Wave from a Vertical Rod, with the Electric and Magnetic Fluxes in Nine Successive Stages During One Semi-Period.

required to illustrate the delivery of one complete magnetic wave into free space. Instead of commencing with the insulated rod just prior to the spark discharge, as in Fig. 7, it is more convenient to commence with the condition indicated at A, Fig. 15, where the electric flux has just completed its movement at light-speed up to the top of the rod, or has climbed up to the summit of the conductor. The flux arrows touching the rod are pointing inwards, indicating a negative charge on the surface of the rod. Ascending electric flux is marked by solid arrows and descending flux by broken or dotted arrows.

Accompanying the upward movement of the converging flux which has culminated in the condition at A, there will be, as previously stated, an associated magnetic flux. The direction of this magnetic flux is indicated by the device of a small circle and a small upright bar, the former on the left-hand of the rod, and the latter on the right hand. The circle may be looked upon as the feathers, or heel, and the small *bar* as the point, or *barb*, of an arrow in a horizontal plane bent into a semicircle about the rod on the side remote from the observer. The device is illustrated in detail in Fig. 15a, where P Q is a vertical section, and *p q* a plan view, of a converging plane of electric flux terminating on the rod E F at its

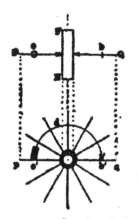

FIG. 15a.—Sectional Elevation and Plan of a Plane Electromagnetic Wave Ascending a Cylindrical Conductor, with Directions of Fluxes. Key Diagram to Fig. 15.

center and sliding upwards from E to F over the rod's surface, like a ring over a peg. When, as shown in Fig. 15a, the small circle *c* is on the left hand, and the bar *b* on the right hand, the ring magnetic flux *c′ d b′* is disposed clockwise about the rod, as viewed from above. Counter clock-wise movement of magnetic flux calls for the circle on the right hand and the bar on the left.

*Rules for Memorizing the Directions of Motion, and of Electric Flux and of Magnetic Flux in Any Single Free Electromagnetic Wave.*

There is a simple law connecting the directions of electric and magnetic fluxes in any simple plane electromagnetic wave. It may be expressed mnemonically in either of the following ways:

(1) Draw an arrow in the direction in which the electric flux points. Let the head of the arrow be supposed to be the head of a man who

runs in the direction in which the flux is running. Then the man's side-extended right hand will point in the direction of the magnetic flux. This rule is illustrated mnemonically in Fig. 16. The

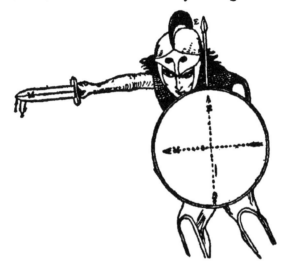

Fig. 16.—Memory Picture for Recalling Directions of Electric and Magnetic Fluxes in Waves.

point of the Greek warrior's sword is supposed to be magnetized and supporting iron nails.

(2) When electric flux, converging upon a conducting rod or cylinder, as in Fig. 15a, points its arrows inwards like the V's in the face of a clock (See Fig. 17), then if the flux is moving—like the light that makes them visible—from the clock towards the observer, the magnetic flux M will point circularly in the direction of the Motion of

the clock hands. Reversing either the electric
arrows E, E, E, or their movement towards the
observer, reverses the direction of the magnetic
flux; but reversing both, leaves the magnetic
flux M pointing clockwise.

### Release of Electromagnetic Wave from Simple Rod Oscillator

Applying either of the rules to the upwardly
moving and inwardly pointing horizontal electric

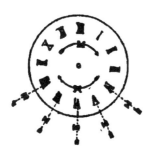

flux on the rod at A, Fig.
15, it will be evident
that the magnetic flux
is directed clockwise, as
viewed from above.

The moment the elec-
tric flux reaches the top
of the rod it is reflected
from the free end. Re-
flection of electric flux
from a *free* end always
requires the direction
of the magnetic flux to
be reversed, leaving the

Fig. 17.—Mnemonic Clock
Diagram of Relative
Directions of Electric
Flux *E* and Magnetic
Flux *M* in an Electro-
magnetic Wave Coming
Towards the Observer
and Surrounding a Con-
ductor at the Center.

electric flux arrows unchanged in direction, but
moving backwards, or retreating. This relation
is a consequence of either of the above rules.
Referring to Fig. 15a, the flux which has reached

the top of the rod at light-speed must keep on moving, and since it can go no further upwards, it commences to descend at light-speed. In Fig. 15, descending electric flux is shown by dotted line arrows and ascending flux by heavy line arrows.

The reversal of the direction of motion of magnetic flux around the rod, in changing from going up to coming down, delivers a blow, by inertia, to the surrounding external ether. In other words, the jerk required to reverse the magnetic flux around the rod between stages A and E of Fig. 15, sets up a counter-jerk or kick in the surrounding ether. The kick sets up two free waves traveling in opposite directions. The electric fluxes in these two waves are mutually opposite; but the magnetic fluxes conspire clockwise. These relations are indicated by the external pairs of arrows at A, which start into existence at the instant of reflection of the central wave from the top of the rod.

At the instant of time represented at B, the second stage, there has been a movement of the flux gliding over the rod, and also a movement in each of the two external free waves. Taking these in order, the gliding flux has commenced to move down at the top, or to double back upon itself, the three lower layers still climbing, but the

leading layer descending. The electric flux arrows point inwards at all parts of the rod, but the
magnetic fluxes are opposed at the top, or tend
to neutralize there. In the external waves, the
ascending one has advanced one stage, and its
wave-front is at *a*. The descending one has
reached the conducting disk, or ground, at the
base, and has been reflected from this surface.
Reflection at a normal conducting surface en-
tails reversal of the electric flux arrows but no
reversal of the magnetic flux arrows. Consequently the lowest layer of the free external descending wave at A has turned its arrows from
outwards to inwards, and the ring is moving
bodily upwards.

At the next stage, indicated at C, the leading
half of the wave on the rod has doubled back on
the following half, the electric flux all pointing
inwards, and the magnetic fluxes completely
neutralizing. There is no resultant magnetic
flux at this instant around the rod. In the free
external waves, the ascending front has reached
b. Half of the external wave which commenced
moving downwards at A, has doubled back upon
itself and is ascending.

Continuing this process to E, we find that the
electric flux slipping on the rod is now all moving
downwards and the magnetic flux is all counter

clockwise. This means in ordinary language that the electric current in the rod is at this instant a maximum in the downward direction. The electric flux arrows all point inwards, so that there is a negative electric charge all over the rod. In the external ether, the direction of motion is altogether upward, with the electric flux inward and the magnetic flux clockwise. The front of the emitted wave has now reached *d*.

A new spark will now cross the air-gap at the base of the rod, not shown in the Figure, and the electric flux will pass over the conducting spark column to the ground at the base. It is reflected back from there with reversal of electric flux arrows, but persistence of magnetic flux; so that there is no kick or disturbance generated this time in external space. At F, the head of the wave conducted to the ground over the rod has turned around and is ascending. The external free wave has reached *e* and is clear of the ground.

At G, there is complete neutralization of electric fluxes, there being no resultant charge on the rod at this instant. But the current wave, as gauged by the conspiring magnetic flux, is at its maximum, and directed upwards.

In the last stage, at I, the flux gliding along the rod has reached its full development in the upward direction, and is about to be reflected back from the free end at the top. This will involve a reversal of magnetic flux and a new shock to the surrounding ether, but in the opposite direction to the shock delivered at A. A new pair of oppositely moving external waves is thereby created, as indicated at I.

After eight more stages have been passed, the external wave will have completely deployed, and the length of this emitted wave will be just four times the length of the rod. The half wave contained between h and i, at I, is twice the length of the rod.

Reviewing the various stages, it will be evident that the electric flux reaches its maximum resultant value near the top of the rod; while the magnetic fluxes, on the contrary, are always in opposition at reflections from the top and reach their maximum near the bottom. This is another way of saying that the electric charge, and electric *voltage* or *potential*, develop maximum amplitude in oscillation at the top of the rod, and the electric *current* at the base.

The diagram of Fig. 15 must not be interpreted too literally. The actual flux distributions are somewhat more complex, and the radius of the emitted wave at I, say, is not exactly equal in all directions to the height h I. The emitted wave becomes sensibly hemispherical, however, after the radius has acquired the length of one-half wave. It is sufficient to observe that there are two sparks for each complete electromagnetic oscillation, and one complete oscillation of electric pressure and current on the rod is accompanied by the emission of one complete free

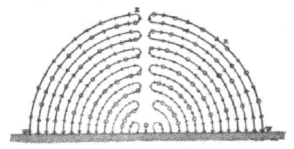

FIG. 18.—Train of Seven Hemispherical Electromagnetic Waves of Decaying Amplitude Liberated by a Rod Oscillator at Center.

hemispherical wave into space. The energy contained in the fluxes of the wave are drawn from the energy of the fluxes oscillating up and down the rod, which are thereby constantly being weakened, and reduced in density.

A diagrammatic vertical cross-section of seven complete hemispherical waves is seen in Fig. 18, as emitted from a simple rod and grounded disk oscillator o at the center. The first wave has attained the radius o w, about 28 rod-lengths from the center; while the seventh wave has just

been released.  The first wave was the most
powerful and is represented in the heaviest lines.
Each successive wave is weaker and weaker.  It
should be remembered, however, that the heavi-
ness of the lines in the illustration only relates to
the strength of each wave at the moment of its
release, or at the moment when it passes a given
point; for as each wave expands, its energy per
cubic meter or cubic foot rapidly diminishes, because the volume occupied by the wave rapidly
increases.  Consequently, the energy per cubic
meter in the first and strongest wave, by the time
it has reached the position w x w, may be much
less than the energy per cubic meter in the shell
of the last and feeblest, but most condensed, wave
at the center, O.

At any point, R, the advance of the wave is in
the radial direction O R, at the speed of light.

The directions of electric and magnetic fluxes
in this train of waves is indicated by the devices
already used in Figs. 15 and 15a.

*Deviation of Electromagnetic Waves from the
    Hemispherical Form Owing to the Curva-
    ture of the Earth*

According to the theory above outlined, the
hemispherical waves of Fig. 18 would travel over
a perfectly flat conducting ground surface at

light-speed and the polar radius o x would be 300,000 kilometers (186,400 miles) long after one second. In practice, when such waves are thrown out by a rod oscillator we have to deal with a moderately conducting spheroidal world surface. This changes the shape of the waves and makes them less geometrically simple. The waves will conform to the curvature of the earth and sea, the successive rings of positive and negative charge running out in all directions over the earth's surface at light-speed, and the feet of the electric flux shells gliding along with them. The waves continue to weaken in intensity per unit of volume as they run, both on account of expanding volume, and owing to sinking into the imperfectly conducting earth surface at their feet, *i.e.*, by frictional dissipation of energy in the ground.

### *Condensation of Fluxes, and of Their Energy, Towards the Equatorial Zone*

Although the direction of movement of these waves is always radially outwards, or perpendicular to the wave front, yet the density of the electric flux in each wave is not uniform all over the surface of the hemisphere. It is greatest near to the ground, and least near to the pole. This is roughly indicated in Fig. 14. Some of the up-

ward electric flux in the inner hemispherical shell
turns back to the earth a short distance above
the surface.   The higher we rise in the shell, the
less flux we find in it, and when we reach the
polar axis Q P all the flux has ceased.   The same
condition necessarily attaches to the associated
magnetic flux in the wave.   The result is that
the energy contained in unit volume of either or
of both fluxes near the ground is greater than it
would be according to simple uniform distribution, by about 60 per cent.   In other words, the
flux densities and energy in any hemispherical
electromagnetic wave are greatest at the ground
or equator and dwindle towards the pole.

### Relations Between Wave-length, Frequency, and Periodic Time

From the relation that the length of the waves
emitted by a simple vertical rod oscillator is four
times the length of the rod, we can readily find
the duration of each wave as it passes any point
on the ground or in the air above.   Let us suppose that the oscillator is kept supplied with
electric energy so as to keep sending out waves
for just one second of time, and that the length
of the rod is, say, 25 meters (27.34 yards).   Then
the length of each wave would be 100 meters
(109.36 yards) measured along any radius.   But

in one second of time, the radius of the outermost wave would have reached to 300,000 kilometers (186,400 miles) and this distance would cover 3,000,000 such wave-lengths. It is clear then that the oscillator must have emitted three millions of waves in the one second of time considered, and also that the time occupied by the wave to pass a given point would be $\frac{1}{3,000,000}$ of a second. This is stated in the customary phraseology by saying that the *frequency* of oscillation is 3,000,000 cycles per second and that the *periodic time* of such waves is $\frac{1}{3,000,000}$ second. In a similar manner, if we know any one of the three quantities wave-length, frequency, or periodic-time of an electromagnetic wave in air, we can instantly assign the other two, because the velocity of propagation is the light-speed in air of 300,000 kilometers (186,400 miles) per second. It is believed that there is hardly any difference between the speed of light in air and in free ether space devoid of air.

# CHAPTER VII

## UNGUIDED, OR SPHERICALLY RADIATED ELECTRO-MAGNETIC WAVES

*Generation of Spherical Electromagnetic Waves by the Discharge of a Double-rod Oscillator*

IF we take a pair of conducting rods A B, C D, and suitably support them insulated in line with each other as indicated in Fig. 19, then on charging the system electrically, with the rod A B, say, positive, the electric flux lines will permeate all the surrounding air in the distribution roughly depicted. So long as the insulation is maintained there will be no magnetic flux. If, however, we raise the electrification to a point at which the air breaks down between the two opposed extremities B C, the electric flux system collapses and runs in towards the spark. At the same time magnetic flux is generated in rings around the axis A D by the inrushing electric flux. The sudden generation of magnetic flux gives a shock to the surrounding ether which sends off a spherical electromagnetic wave into surrounding space at light-speed.

64

The contours of five successive spherical waves is given diagrammatically in Fig. 20, with reference to the pair of discharging rods at o, the center of disturbance. If the two electrified rods

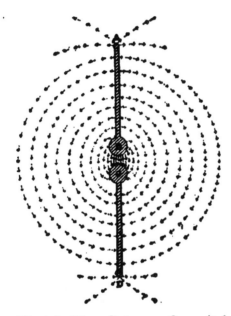

FIG 19.—Electric Flux Between Oppositely Charged Conducting Rods.

have no source of energy to maintain their oscillations except the original charge, the successive outgoing waves carry that energy away, while some of the remainder is dissipated in heat in the spark and also in the surfaces of the rods, so that the oscillations rapidly die away. It is not impossible, however, to supply electric energy to

the rods as fast at it is radiated externally and
dissipated locally, so as to maintain the oscillations indefinitely, although it is very difficult to
do this experimentally. In such a case, the wave

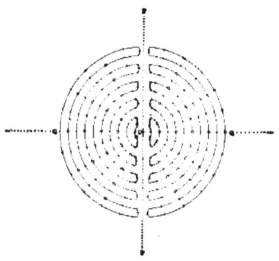

FIG. 20.—Diagram of Section through the Polar Axis
of a Train of Five Spherical Electromagnetic
Waves Emitted by a Double-Rod Oscillator at the
Center.

train would go on extending and expanding indefinitely at light-speed. If the rods could somehow be placed in free space, remote from the
earth and all conductors, the spherical waves
would keep on moving radially outwards. In
practice, on this earth, the waves must almost
immediately strike the surface of the ground on
one side at least, and be reflected there, not to

speak of the influence of neighboring walls, trees, etc., so that the pure spherical form cannot be maintained.

Figure 20 shows that the polar axis P P is in the line of the rods at the center o. On this axis the electromagnetic fluxes and energies disappear. At and near the equator Q Q, the fluxes are densest and their energies are a maximum, for any given radial distance from the center. The length of each wave is four times the length of either rod, as in the hemispherical waves considered in the last chapter; or it is twice the length of the double-rod oscillator A D of Fig. 19.

## *Spherical Electromagnetic Waves Identical with Long-wave Polarized Light*

Physicists are now agreed that such an oscillator as above described, if kept supplied with energy for radiating electromagnetic waves, would emit *light*. That is to say, such electromagnetic waves constitute light; although not ordinary light such as is recognized by the eye. The main difference between such waves and ordinary light lies in the wave-length. The human eye is able to recognize as light electromagnetic waves whose length lies between the limits of 0.4 micron* ($\frac{1}{64,000}$ inch), in violet light, and 0.8 micron ($\frac{1}{80,000}$ inch) in red light. Electromagnetic waves which are either shorter or longer than this are not directly visible; although they may still be objectively regarded as light.

## Modern Electric Theory That All Matter is Ultimately Electricity

Ordinary matter, such as a piece of match-wood, is believed to be made up of ultimate particles called molecules, too small to be seen by the microscope. Molecules are chemical combinations or chemical groups of atoms. Atoms are the supposed ultimate particles of elementary substances, or the smallest pieces of such elementary substances which can exist separately as such. Atoms in their turn are now supposed to be each constructed of much more minute or ultra-ultra-microscopic electrical charges, called *electrons*, there being a definite number and organization of electrons to each atom of an elementary chemical substance. Consequently, all the matter in the universe is ultimately constructed, according to this theory, of definitely

* The micron is the term used in microscopy for the one-millionth part of one meter from the Greek *mikros*, small. It is usually designated by the Greek letter mu $\mu$.

organized electric charges, or of electricity. An atom of hydrogen, for example, is supposed to comprise about 800 electrons in some definite organized orbital or planetary movements, an atom of oxygen about 10,000 electrons in a different grouping of orbits; and so on, for other elementary substances.

When atoms are heated, as for example, the hydrogen and carbon atoms of wood, by setting fire to a match, the electric charges or electrons within the atoms are regarded as being forced into rapid and violent oscillation, whereby electromagnetic waves are radiated off. Since these atomic oscillators are of ultra-microscopic dimensions, so too are the lengths of some of their electromagnetic waves. Those waves whose lengths lie between 0.4 and 0.8 micron are perceived by our eyes as light. A pair of little rod oscillators, as in Fig. 19, each about 0.2 micron long ($\frac{1}{120,000}$ inch) excited into sustained radiation, would give off waves of red light, the longest waves by which the retina of the eye is affected.

### Virtual Ultra-microscopic Oscillators in Heated Matter and Their Emitted Waves

Looked at in another way, the shortest electromagnetic waves that have yet been produced by the discharge of electric rods or spheres are a few centimeters, or inches, in length. In order to make visible light in the same manner, we should have to use ultra-microscopic particles as discharging bodies. On the other hand, the waves employed in wireless telegraphy usually vary between 100 meters (109.4 yards) and 10,000 meters (6.21 miles) in length. The latter would include in one wave length 25,000,000,000 waves of violet light, the shortest detected by the human eye.

The velocities of all electromagnetic waves being apparently the same, whatever their length, their frequencies differ in a similar range. The frequency of a 10-kilometer (6.21 miles) wave would be 30,000 cycles per second, or each oscillation would occupy $\frac{1}{30,000}$ second in execution. But the frequency of violet light is 760 millions of millions per second. Each color of the spectrum has its own frequency and corresponding wave-length.

There is one other difference between visible light and spherical electromagnetic waves produced by electric discharge between conductors as in Fig. 19. This is in regard to the directions of the poles of the waves. In Fig. 20, the polar axis of the waves always lies in the line of the rods, no matter how far the waves may extend into space. There is no energy emitted along this axis. Let us suppose that the rods are ultra-microscopic, so that they are enabled to emit waves short enough to affect the eye, and that their energy of oscillation is somehow sustained. Then the point *o* in Fig. 20 would be a luminous point, shining with one particular color of the spectrum. The brightness of the point would, however, be greatest in the equatorial plane Q Q, and it would dwindle to zero as we moved the eye to the polar axis. An eye at Q would see the shining point; but an eye at P would see nothing. This is contrary to experience with glowing material points. A lighted match or glowing point, sends out rays in all directions. '

The discrepancy is accounted for by the fact that the polar axis of the electric disturbance in a luminous point, supposed to be due to oscillating electrons, is constantly shifting its direction in space. One wave may have its polar axis vertical, but after a few more have passed by, there will be a wave with its axis horizontal, and later again the axis will be vertical; so that in a single second of time including millions of mil-lions of waves, the atomic electric oscillators will have turned into all directions and will have made many gyrations. Consequently the eye will have received in that time many waves in their equatorial zone and also many in their polar zone, so that the average effect will be the same in every direction. We need only suppose the rod oscillator of Figs. 10 and 20 rotated about the spark center in all directions at great speed during the emission of a long train of waves, to see, in imagination, the effect that would be produced upon the eye of an observer at any distant point.

Electromagnetic waves or light waves in which the polar axis remains fixed in space, as in Fig. 20, are called *plane-polarized* waves. The *plane of polarization* is the equatorial plane Q O Q, parallel to which all the magnetic flux-paths are disposed. Ordinary, or non-polarized, light may be artificially plane-polarized by optical methods.

We may say then that ordinary visible light consists of electromagnetic waves of sustained amplitude—*i.e.*, not merely a few decaying oscillations—within a certain sharply limited range of small wave-lengths or high frequencies, and with the polar axis in all directions in rapid succession. Ordinary daylight contains almost all the wave-lengths within the visible range, showing that vast numbers of atomic oscillators of different "lengths" are simultaneously operating in the glowing solar surface and are mingling or superposing their electromagnetic waves. These waves reach us in about 500 seconds after they leave the atomic electric oscillators in the solar atmosphere.

### Solar Wireless Telegraph Waves, in Broad Sense, Necessary to Life of Human Beings

In a certain sense, therefore, every shining star in the heavens is constantly sending out spherical electromagnetic waves within the range of visual perception, besides probably many longer waves, outside of that range. In this particular sense we are constantly receiving wireless telegraph waves from every visible orb, and the message received is not news but light. Moreover, since all animal energy is derived from plants, and all plants build up their substance from the energy contained in the sunlight they receive, it follows that all our muscular energy is derived indirectly from wireless telegraph waves received from the sun.

### Union of Optics with Electromagnetics

All of the phenomena of light, reflection, refraction, polarization, interference, etc., which have been within the special study of Optics for many decades, have in recent years been imitated, on a relatively large scale, by electromagnetic waves set up by the discharge of elec-

trified conductors. In fact, a few of the properties of optical waves which are difficult to detect,
by reason of the excessively short optical wave-
length, are more easily studied and revealed in
electromagnetic waves in the electrical laboratory.

## Classification of Types of Electromagnetic Waves

Summing up the conclusions reached in the
last few chapters, we may say that discharges
between two rods or conductors set up spherical
waves. Discharges between a conductor and a
plane conducting surface, such as the ground
approximates, set up hemispherical waves.
Waves guided between a pair of parallel wires;
and between an aerial wire and the ground are
cylindrical waves, moving end-wise. The waves
employed in ordinary wireless telegraphy are
initially hemispherical waves conforming to, or
guided by, the spheroidal earth.

# CHAPTER VIII

## PLANE ELECTROMAGNETIC WAVES

### Hemispherical Waves of Large Radius Are Virtually Plane at Any One Point

ANY small section or piece cut from the front of a hemispherical wave is practically flat, or plane, when the wave is remote from its origin, just as the earth's spherical surface is practically flat, or plane, at any one point on the ocean, because its radius is relatively so large. Consequently, any hemispherical wave advancing over the surface of the earth or sea may be regarded as plane locally. It comes along like an invisible upright wall.

A section of a single such wave is shown in Fig. 21, taken along the line of march V V. G G represents the surface of the ground. The electric flux rises perpendicularly at P P, or very nearly so. If the earth conducted perfectly, the electric flux would rise strictly vertical from it. Imperfect conductivity causes a wave to lean over, or bend forward slightly, as it moves, so that a perpendicular to the wave front would no longer lie parallel to the ground but would point into it. For practical purposes, however, we may take the electric flux as perpendicular.

In the front half of the wave, we have taken the flux P P as pointing upwards, corresponding to a moving positive charge on the ground beneath; while in the rear half these conditions are reversed. This relation of directions depends upon the direction of the fluxes in the oscillator at the moment that this particular wave was born.

The amplitude of the current waves on the ground are indicated by the curved line f p o n r. The line of zero current is the line *z f o r z*. The direction of the magnetic fluxes is also indicated, by circles where the flux is directed away from the observer and by short horizontal bars where towards the observer.

The wave front has reached F F, while the rear of the wave is at R R. The wave-length is, therefore, the horizontal distance F R. At the central vertical plane O O, midway between the positive and negative developments of the wave, the fluxes are zero and their energies are consequently zero. The fluxes are densest at the central planes P P and N N, and their energies in a given volume are also a maximum at these planes.

In practice, there will usually be a train of successive waves moving over the ground in place of the solitary wave of Fig. 21.  A wave

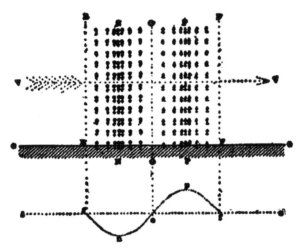

FIG. 21.—Section of a Single Electromagnetic Wave Along Line of Advance and Near to Surface of the Ground.

train in wireless telegraphy does not usually contain many waves, and their amplitude successively diminishes; so that the final waves in the train are extremely feeble.

*Analysis of a Single Wireless Telegraph Wave At and Near the Earth*

A section of the wave in the plane of P P, Fig. 21, is given diagrammatically in Fig. 22.  It appears as a number of parallel equidistant, ver-

tically rising, electric flux lines, crossed at right
angles by a number of parallel equidistant horizontal magnetic flux lines.  This means that both
the electric and the magnetic fluxes have uniform
intensity in this plane.   The charge moving upon

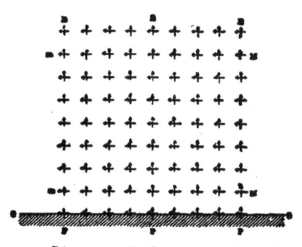

Fig. 22.—Diagrammatic Section of Plane Vertical
    Electromagnetic Wave Parallel to Wave-Front
    and Advancing Towards Observer, with Electric
    Flux Rising Vertically from Positive Charge on
    Ground and Magnetic Flux Horizontal.

the surface of the ground below the wave is positive.   Since the ground is not a perfect conductor, the fluxes penetrate into it to some extent.
This causes a certain amount of energy to be
expended in the penetrated layer of soil as heat,
derived from *eddy currents*, of parasitic electric
currents, in the soil.  The energy expended at

the foot of the wave has to be paid for from the stock of energy residing in the wave as a whole; so that energy is fed downwards as the wave runs along, causing a weakening of the moving fluxes, in addition to the weakening caused by the simple hemispherical expansion of the wave.

### Electric and Magnetic Forces Embodied in the Wave and Moving Therewith

If we could compel the wave to stand still for a moment, instead of running by the observer at light-speed, we should expect to find that a positively electrified pith ball would be urged upwards by the upwardly pointing electric flux of the wave as depicted in Fig. 22; while a delicately poised magnetic compass needle would tend to align itself along the lines M m in the wave front. In any ordinary wave, however, these electric and magnetic forces would be of very feeble magnitude. The fact that they are able to produce recognizable effects as they pass by is due to their enormous speed, the speed of light in the medium.

### Transparency of Electric Non-conducting Obstacles to Long-wave Light

When a plane electromagnetic wave, many meters or yards long, running along the surface

of the ground, strikes a brick wall, or a wooden·
frame house devoid of metal, it passes through
these obstacles with very little disturbance.  This
means that if our eyes were capable of responding
to these waves, so that they produced the sensation
of some type of color, such non-conducting obstacles would be transparent to that color of light,
and we could look through a brick house or a
wooden house without difficulty, when objects
were illuminated by these waves, or in this
imaginary type of color.  If, however, the advancing waves strike an electrically conducting
obstacle, such, for example, as a simple vertical
metallic rod, indicated in Fig. 23, the obstacle
will either absorb or reflect the waves and will
cast a shadow beyond it, the shadow being, of
course, invisible to us; since the waves are invisible; but the shadow can be determined and
mapped out by suitable electric apparatus; or
by what may be called an artificial eye.

### Shadow Cast by a Vertical Electric Conductor in the Path of an Electromagnetic Beam

The electric flux only is indicated in Fig. 23
advancing from left to right, over the ground
G G.  At *a*, it is about to strike the vertical
metallic rod A B, connected with the ground.
Such a vertical might be a leaden water-pipe, or

a copper wire.  At *b* the wave has passed the
vertical and a gap has been thereby torn in the
wave.  The lower edge of the wave at the rent is
bent backwards and the subsequent direction of
movement of this edge, being always perpendicu-.
lar to the local surface, is downwards as well as

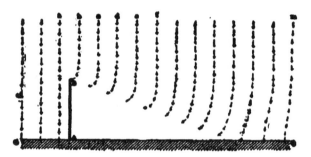

FIG.  23.—Diagram Indicating the Electromagnetic
     Shadow Cast by a Vertical Conductor in the Path
     of an Advancing Plane Wave.

to the right.  In the successive positions *c d e f*,
the wave is spreading down from above, and at
*m* the rent in the wave net has been repaired, or
the shadow behind A B has been filled up, the
energy of the flux put into the patch being drawn
from the remainder of the wave.  It is to be
understood, however, that the diagram of Fig. 23
makes no pretensions to geometrical accuracy,
and the exact contour of the shadow thrown by
a conducting obstacle is not yet determinable
with precision.

After the wave has struck the rod A B, a disturbance is reflected back from the rod into the region A B n G, at the same time that the shadow

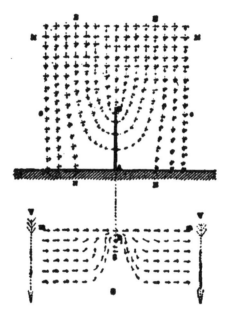

Fig. 24.—Diagrammatic Sections in Elevation and Plan of Electromagnetic Wave Striking a Vertical Conductor while Advancing Towards Observer.

is cast beyond. This reflected disturbance is, however, omitted from the illustration.

Fig. 24 presents a sketch, both in elevation and in plan, of the actions occurring when the wave strikes the vertical conductor. The wave is supposed to be advancing towards the observer. It will be seen that the electric flux, which is every-

where distributed as in Fig. 22 (reversed), before
it strikes the vertical A B, is drawn in on each
side to the rod and converges on the same, continuing to run down the rod for a little while
after the wave has passed.   The magnetic flux
is shown in the plan at the base of the illustration.
Before the wave reaches the rod, as at m m, the
magnetic flux lies in a horizontal straight line,
parallel to the wave front.   As soon as the wave
strikes the rod, the magnetic flux bends around
it clockwise, and also descends the rod at light-
speed.   S S is the shadow cast by the rod B, or
the space denuded of magnetic flux for a certain
distance behind B.

Looking at the action from another standpoint,
we may, in the light of our provisional electromagnetic theory, consider that the electric flux
advancing over the ground brings a local right-
handed torsional stress upon the ether, which,
by electric rigidity, resists the stress and limits
the flow of ether to that small amount found in
the wave front as indicated by the horizontal
magnetic flux lines M M, Fig. 24.   At soon as
the electric flux strikes the conducting rod A B,
the elastic rigidity of the ether is lost, owing to
the action of the conductor and the electrons
residing in it.   The ether at the rod gives way
before the stress, and flows bodily around the rod

in dense magnetic flux streams. On such a
hypothesis, a conductor behaves like a gap in the
ether, and the advancing electromagnetic wave
pours electric and magnetic fluxes spirally or
vortically down into the gap as it goes by.

### Comparisons Between Reflection of Short and Long Waves of Light

Whatever hypothesis we adopt to assist the
mind's eye in depicting the process, we must
expect to find the action similar to that which
occurs when half-micron electromagnetic waves,
*i.e.*, visible light, strike an opaque obstacle.
There is a reflected wave train thrown back by
the obstacle. There is also a shadow cast behind it, and there is energy absorbed into the
substance of the obstacle. The width of the
shadow cast by a parallel beam of light is apparently no wider than the obstacle; whereas in
Fig. 24, the shadow cast is indicated as being
many times the width of the vertical rod. But
it has to be remembered that if the optical
shadow of a rod were one quarter of a wave-
length wider than the rod, the difference would
be only about a sixth of a micron ($\frac{1}{150,000}$)
and quite insignificant; whereas if the rod A B
had the same height as the simple rod oscillator
which originally emitted the wave, a shadow hav-

ing a breadth of a quarter wave-length, would be as broad as the height A B, or the distance N N in Fig. 24, and would be equal to the height A B, a very appreciable distance.

### Gashes Torn in Electromagnetic Waves by Vertical Conductors on the Earth

It is evident that upright metallic rods, such as lightning-conductors, tear rents in any passing electromagnetic wave running along the ground. On the other hand, a conductor parallel to the ground, such as a trolley-wire, or an overhead telegraph wire, does not sensibly affect a passing electromagnetic wave. Looked at in another way, a vertical rod is cut by the rushing horizontal magnetic flux at light-speed, and acts like a single-wire dynamo, moving through a very feeble magnetic field at the speed of light. Again, a vertical rod picks up a certain difference of electric potential between the electric flux at its top and at its base. In either of these ways, the rod becomes the seat of an electric impulse or *electromotive-force* during the brief interval in which the wave is passing by it. But if we place the rod horizontal, instead of vertical, the electric flux in the wave will cut the rod perpendicularly and the magnetic flux, in cutting, only acts upon the thickness of the rod; so that the electromo-

tive force set up therein by the passing wave will
be insignificantly small, and will be directed
transversely or across the diameter of the horizontal rod.

Accordingly, when a single electromagnetic
wave hits a vertical rod, a rent is torn in the wave,
and the breadth of the rent, although not yet
accurately known, may, perhaps, be a quarter
of a wave-length. The energy which resided in
the wave within the region torn out, is available
for setting up electric currents in the rod, after
allowing for what is lost by reflection and secondary radiation.

It may be readily imagined that when an
electromagnetic wave strikes a steel bridge, or a
steel sky-scraper office-building, it casts a long
shadow, and a relatively large quantity of energy
is torn out of the wave. Trees also, and shrubs
too, in lesser degree, have been found to be feebly
conducting, and it is believed that they absorb
energy from waves passing them. This fact
taken in connection with the imperfect conductivity of dry soil, in comparison with sea water,
accounts for the considerably greater distance at
which electromagnetic waves can be transmitted
and detected over the ocean than over land. The
signaling distance range over the sea is, roughly,
double the signaling distance range across country.

*Elementary Analysis of Electric Oscillations Set
Up in a Vertical Conductor by the Passage of
Waves*

It is important to notice the principal events
that occur in the neighborhood of the vertical
rod after it has been struck by the onrushing
electric wave. Fig. 25 indicates diagrammatically nine successive stages in half a complete
cycle of these events. The line of crossed arrows immediately under the letters A B C
. . . . I, represents the directions of electric
and magnetic flux in the advancing wave over
the rod. Thus at A, the conditions are those
indicated in Figs. 23 and 24; namely, the electric
flux is pointing downwards and the magnetic
flux pointing to the right, as viewed by an observer who sees the wave advancing towards him.
At E and F these fluxes are in the act of reversing
through zero, corresponding to a plane such as
O O in Fig. 21. At I, the fluxes have completely
reversed.

Underneath each diagram of a rod section in
Fig. 25, there appears a plan view showing the
direction of magnetic flux in the wave just before
striking the rod, and also of magnetic flux encircling the rod. Thus, at A, the magnetic flux in
the air is at full development towards the right

hand of the observer, while around the rod it is clockwise. At C, the clockwise magnetic flux encircling the rod has reached full development, or the electric current over it is a maximum. Between E and F the magnetic flux in the air reverses or passes through zero. At G, the magnetic flux encircling the rod passes through zero. At I, the magnetic flux in the air-wave has developed completely in its reverse, or left-handed direction.

Examining the rod at A, it will be seen that the electric flux of the passing wave has converged upon it, as already seen in Figs. 23 and 24. This flux immediately starts to run down the rod to ground, as indicated by the long dotted arrow. The instant it begins to run, the electric flux reverses direction, or assumes the outward direction shown at B, the magnetic flux remaining clockwise, as viewed from above. As soon as the flux reaches the conducting ground at the base of the rod, it is reflected thence upward, with a new reversal of electric, and maintenance of magnetic, flux direction. At E, the stream of flux on the rod is about to reach the top. At the top, the flux reverses magnetically, or is reflected downward, with persistence of inward electric flux. At G, the magnetic flux is half clockwise and half counter-clockwise, representing zero of

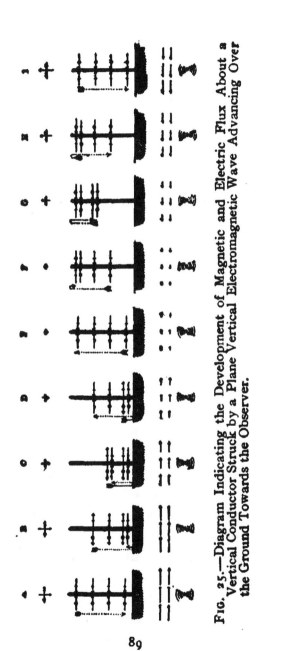

FIG. 25.—Diagram Indicating the Development of Magnetic and Electric Flux About a Vertical Conductor Struck by a Plane Vertical Electromagnetic Wave Advancing Over the Ground Towards the Observer.

89

current, but maximum electric potential. At l,
the flux is in full descent again, with counter-
clockwise magnetic field.

## Resonance in Electric Conductors Struck by Wave-trains

We have purposely chosen the length of the
rod as one quarter of the length of the plane wave
advancing through the air. This brings about
such an adjustment of the motion of flux over
the rod as enables the next succeeding wave to
add to, or increase, the movement. If we extended the diagram of Fig. 25 through eight more
such phases we should return to the original condition at A, when the flux in the next wave would
not only repeat the cycle, but would also increase
the amplitude. If the rod conducted perfectly,
and also the ground at its base, each wave as it
arrived through the air would add to the fluxes
running up and down the rod, on the familiar
principle of the child's swing, whose oscillations
may be increased by timing the pushes to the
natural period of oscillation. In this case, however, the swing of the rod is adjusted by its length,
so as to be in rhythm to the train of arriving
waves. Such a condition of coincidence between
the times of arrival of the successive wave-crests,

and the natural time of electric oscillation of the rod, is called *electric resonance.*

If we could obtain a very long train of uniform advancing waves and adjust the length of the vertical rod into resonance therewith, retaining perfect conduction, the fluxes running up and down would increase indefinitely, were it not for secondary radiation. That is, the rod, excited in this way by arriving waves, would become an oscillator in its turn, and discharge the energy it received in a new series of radiating waves, as in Fig. 15. In practice, however, the waves received through the air have such feeble amplitude, they decay so soon, the number in a train is so small, and the conductivity of the rod and ground base is so far from being perfect, that even when the rod length is adjusted into resonance, the currents developed over the rod, as in Fig. 25, are comparatively feeble. The secondary radiation is, therefore, insignificant.

If the length of the rod is in the opposite condition to that required for resonance, the fluxes generated thereon by the first wave will be opposed, instead of aided, by the fluxes generated in the second, and so on. Consequently, there will be comparatively feeble currents set up on the rod. If, however, the length of the rod is adjusted for resonance, there will be a building

up of electric current on the rod, unless the arriving wave train is too short, or unless the electric
obstruction and want of conductivity in rod and
ground suppress the development.

In order to adjust the rod to the resonant condition, it is not always necessary to alter its height.
The virtual length can be altered by the insertion
of a suitable form of conductor or wire at the
base, between rod and ground, in a manner to be
described later.  In such a manner the time of
oscillation of a rod can be altered without changing the actual height.

### Résumé of Conditions Attending the Impact of Waves Against Vertical Conductors

Summing up the above results, we find that a
vertical conductor connected with good conducting ground, and set up anywhere in the path of a
train of electromagnetic waves, will have alternating, or to-and-fro electric currents set up on it,
the energy contained in these currents being the
energy in the up-and-down moving fluxes, which
energy is drawn from, or scooped out of, the
arriving electromagnetic waves as they pass by.
These alternating currents in the rod are capable
of being built up, or successively increased in
strength, by the impulses of the successive waves,
if there be resonance, i.e., if the natural time of

oscillation of the rod be the same as the periodic time of the arriving waves. For a simple vertical rod, devoid of inserted apparatus, this will be when its height is one quarter of the wave-length, and therefore equal to the height of the simple vertical rod oscillator which is capable of originating such waves. In other words, if the arriving waves have been produced by a distant simple rod oscillator, resonance will require the heights of the oscillator and of the receiver verticals to be equal. Resonance would be capable, theoretically, of setting up an indefinitely great amplitude in a perfectly conducting receiver rod, set in perfectly conducting ground, with a constantly maintained succession of waves at the oscillator, were it not for secondary radiation of waves from the receiver. In practice, however, the alternating currents set up at the receiver may be materially increased by bringing the receiving rod into resonance, but the development is arrested by imperfect conduction at the receiver, as well as by discontinuity of the oscillations, or small trains of waves at the oscillator. Moreover, the insertion of a receiving instrument into the path of the vertical receiver rod also causes energy to be absorbed, and interferes with the production of resonant increase of oscillations.

*Energy of Electric Oscillations, or Oscillating
Currents, Set Up in a Vertical Receiver*

The energy which is available for producing
such electric currents by any wave at the receiver
depends upon the energy in the entire hemispherical wave at that moment.   It will evidently be
but a very small fraction of the total energy of the
wave, when the receiver is far from the oscillator,
since the area of the wave which can come into
contact with the receiving rod, or into its region
of influence, is so small.   If we suppose that the
receiver has a height of a quarter wave-length,
for resonance, and that the effective breadth of
area from which energy is drawn, as in Fig. 23,
is also a quarter wave-length, then the fractional
part of the wave's energy available for producing
electric current at the receiver, is the square of
the height of the receiver rod, divided by the
superficial area of the hemisphere occupied by
the entire wave at the instant it strikes the rod.
For example, if we suppose that a certain wave
in a series emitted by an oscillator contains at the
moment of shipment an amount of energy equal
to 1 kilogramme-meter (7.24 foot-pounds, or the
work done in lifting one pound through a vertical
height of 7.24 feet), then a quarter-wave vertical
receiving rod at a distance of 30 kilometers

(18.6 miles), with a height of say 31.6 meters (103.6 feet) might perhaps absorb energy from the wave as it passed, over a height of 31.6 meters and a breadth of 31.6 meters, or an area of wave surface amounting to 1,000 square meters (10,760 square feet). But, neglecting the curvature of the earth, the area of a hemisphere 30 kilometers (18.6 miles) in radius would be 5,655 millions of square meters (60,800 millions of square feet); so that the electromagnetic energy capable of being drawn on to the rod would be $\frac{1000}{5,655,000,000} = \frac{1}{5,655,000}$ of a kilogramme-meter, or 17 ergs. The received energy should be about 60 per cent. greater than this, because of the greater density of flux and energy in the equatorial zone of the transmitted wave, *i.e.*, near the earth's surface. On the other hand, however, a distinct reduction would have to be made for the energy wasted in transmission along the surface of the soil, by reason of the earth's imperfect conductivity, or for other vertical conductors, such as metallic structures, or trees, intervening between oscillator and receiver.

Our knowledge is still very imperfect as to the effective surface area drawn upon by a vertical rod, and also as to the amount of energy drained from the feet of an advancing hemispherical wave by reason of the earth's imperfect conductivity.

This loss is known to be greater after dry weather than after rain.  It is, however, clear that according to the working theory outlined above, the energy capable of being received by any such ver-vertical rod increases as the square of its height, assuming that the resonant condition is maintained, and also inversely as the square of the distance between the sending and receiving stations.  The total energy available for producing alternating electric currents at the receiver will be the sum of the successive fractional amounts drawn from each single wave in turn, assuming that the successive effects can be prevented from canceling or annulling each other, by the adjustment of the receiver to the resonant condition.

# CHAPTER IX

## THE SIMPLE ANTENNA OR VERTICAL ROD OSCILLATOR

### *The Antenna and Transmitting Apparatus*

WE have already arrived at the conclusion from preceding chapters that wireless telegraphy ordinarily employs hemispherical electromagnetic waves emitted from a vertical rod oscillator or antenna, in short trains of from two to thirty waves of successively diminishing amplitude. On or near the ground, or sea level, and at a great distance from the transmitting station, these waves are for all practical purposes plane waves, advancing over the conducting ground, or sea, with the speed of light, in a direction radial to the sending station, after allowing for the curvature of the earth. We now proceed to consider the essential elements of the antenna, and of the apparatus employed, at the transmitting station.

The simplest type of vertical antenna or rod oscillator is represented in Fig. 26. It consists essentially of a single vertical metallic wire A B,

suspended from an insulator I, which is supported from a wooden mast structure indicated
in dotted lines. This vertical wire, air-wire,
aerial, or antenna, is insulated from the ground

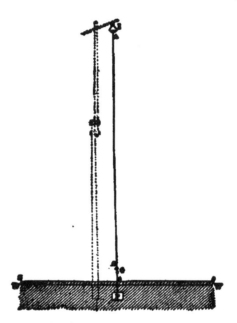

FIG. 26.—Essential Elements of a Simple Vertical Oscillator, or Antenna, for Emitting Hemispherical
Electromagnetic Waves.

by the air-gap G, so long as it is electrically in-
active. The lower terminal of the air-gap com-
municates with a metallic plate P sunk in moist
earth, or below low-tide level, if on the seashore.
Sometimes bare wires w w are laid out radially
from the ground wire in various directions, at or

near the surface of the ground, so as to improve the local conductivity of the soil, and help to form a good electric mirror at the ground surface *s s s*, from which the waves may be reflected back and up the antenna, not only at the conductor, but in its vicinity, as indicated in Fig. 15.

The length of the wave emitted by a simple vertical wire antenna as shown in Fig. 26 is believed to be very closely four times the height of the antenna A B G S. Thus, if the antenna had a height of 30 meters (32.8 yards), above perfect ground, the length of the waves sent out would be 120 meters (131.2 yards). The number of such waves which would cover the distance travelled by light in one second would be $\frac{800,000,000}{120} =$ 2,500,000; so that there would be two and a half millions of such waves occupying one second, if the oscillator could be kept at work for that time. This means that the frequency of the waves would be 2,500,000 cycles per second, or the time occupied by any one complete wave to pass a given point would be $\frac{1}{2,500,000}$th second. If we call the one millionth part of a second one *microsecond* for convenience of description, then one complete wave would pass off in $\frac{1}{2.5}$ microsecond. Since each wave contains both a positive and a negative impulse, either impulse would pass by in $\frac{1}{5}$ of a microsecond.

## *The Large Activity of an Antenna*

Owing to this extremely short period of oscillation, antennas are remarkable for their activity
or power.   The amount of energy which can be
stowed away in a simple vertical antenna as
electric-flux energy in the surrounding ether, by
charging it to a suitably high voltage, is comparatively small, being usually not more than 20
gramme-meters (0.14 foot-pound), or the work
done in lifting 20 grammes to a height of one
meter.   When this energy is released, by the
discharge of the antenna across the spark-gap
G, Fig. 26, part of this energy is expended in the
heat of the spark and in heating the surface of
the conducting antenna.   The remainder is
available for radiation as a series of electromagnetic waves.   Perhaps not more than 3
gramme-meters (0.022 foot-pound) of energy will
be shipped off in any single wave.   Nevertheless,
this energy is shot off by this particular antenna
in $\frac{1}{2.5}$ part of a microsecond, and the average
rate of power radiation during this brief interval
will thus be 7,500,000 gram-meters per second,
or 7,500 kilogramme-meters per second, or about
100 horse-power.

With the aid of auxiliary apparatus, an antenna
may be capable of radiating electromagnetic wave

energy at the rate of hundreds of horse-power;
but only for a few microseconds at a time, so
that its average power in one second, or in one
minute, during its operation, may be only a
small fraction of a horse-power.   An antenna of
the simple type shown in Fig. 26, looks like a
very simple and innocent machine;  but, when
thrown into electric vibrations, it may throw out
as much power as it takes to operate a high-
speed electric locomotive;  only it does not keep
the power up.   The case is somewhat similar
to that of a revolver, which is being fired, say,
three times per second.   At each explosion the
power of the machine is relatively very great;
but between shots the power falls to nil;  so that
the average for one second, the power of the
machine, or its mean rate of throwing energy off,
is comparatively low.

In practice, single-wire antennas are seldom
used, and multiple-wire antennas are customary.
The purpose of employing a plurality of conductors is two-fold.   In the first place, the larger
surface of the antenna gives more electric flux
in the air when charged, and this increases the
stock of energy held by the antenna prior to release and radiation.   In the second place, the
larger surface permits of a more free active radiation or discharge of electromagnetic waves into

the surrounding air, independently of the amount of energy to be released.

## Cylindrical Antennas

Fig. 27 represents the cylindrical type of vertical antenna, one having four parallel vertical

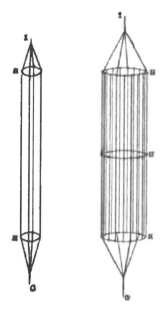

wires and the other sev-
enteen. Any convenient
number may be used.
The wires are usually
soldered to two or more
metallic hoops H H,
which not only strength-
en the structure mechan-
ically, but also keep the
oscillations symmetrical
electrically. The diam-
eter of these hoops may
range from 30 centime-
ters to several meters (1
foot to several yards). If
the component vertical
wires are not further
than, say, half a meter

FIG. 27.—Types of Cylin-
drical Frame Vertical
Oscillators.

apart (19.7 inches) these bird-cage cylinders are
almost equivalent, electrically, to complete sheet-
cylinders of metal. The bird-cage cylinder of
multiple parallel wires is of course far superior

mechanically to a complete sheet-cylinder or large pipe, both in cost, lightness and freedom from wind-pressures. These cylindrical metallic frames may be supported by suitable insulators from a mast-arm at I, their lower extremities G leading to spark gaps.

In some instances a cylindrical antenna is formed of a rigid vertical steel tube, bolted together in sections and supported on insulators at the base. The tube or cylinder is prevented from falling by guy-ropes running in various directions, and in which insulators called strain-insulators, because they are subjected to tension, are inserted at some suitable point or points.

### Harp, Fan and Inverted Cone Antennas

Other forms of antenna in use are outlined in Figs. 28 and 29. The former indicates the harp type. This is conveniently supported between two wooden masts, as indicated in dotted lines, but a single mast may serve if the wires are sus-. pended from a horizontal arm. The five wires shown are connected at the top and at the bottom by horizontal wires. The entire conducting frame is supported by insulators at I I I I. The harp is connected to the spark-gap by the wire G.

The fan-shaped antenna of Fig. 29 is sometimes used on board ship. The stout wire I I is

strung between the masts B A and D C, being
supported by end-insulators I I.   The descend-

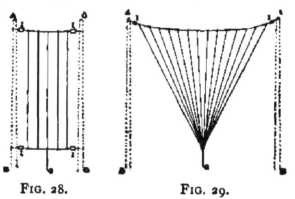

FIG. 28.                 FIG. 29.

FIGS. 28 and 29.—Types of Harp-Shaped and Fan-
Shaped Antennas.

ing wires are each connected to the top-wire I I
above and to the central point below; whence a

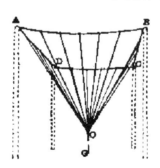

wire G runs to the spark-
gap and ground, or on a
steamer to the metallic
frame of the hull.

At some stations a ser-
ies of fans are connected
together into an inverted
cone, as seen in Fig. 30.

FIG. 30.—Type of Inverted Here four masts support
Cone Antenna.

a metallic rectangle,
through insulators not shown in the diagram.
Metallic wires drop from the rectangle at inter-

vals to the central point O, whence a wire runs across a spark-gap to ground.

Whatever the form of the antenna, cylindrical, harp-shaped, fan-shaped or conical, the object sought, already mentioned, is to increase the electric flux, and electric energy associated therewith, in the charge of the antenna, and also to facilitate the emission of the waves into space at the recoils from the upper end or ends of the antenna.

### Electric Oscillations on Antennas Skin Deep

The thickness of the individual wires forming the antenna is of secondary importance. It is the surface of the wires which is of principal consideration. The high-frequency electric currents, or oscillations, running up and down the antenna, are not able to penetrate below a certain skin depth into the conductor, say 1 mm. The higher the frequency, the less the penetration, and the thinner the effective conducting skin. The wires are usually of copper, and about 4 mm ($\frac{1}{8}$ inch) in diameter.

Other things being equal, the higher the antenna, of whatever form, the more electric flux, charge, and energy it will hold; so that the power it can release is greater. At the same time the length of the wave tends to be greater.

### Sources of Energy for Feeding to an Antenna

The source of electric energy for charging the antenna is generally an induction coil, or spark coil, excited either by a dynamo, or by a voltaic battery. If a voltaic battery is used, it is commonly a secondary, or storage battery, charged by, and receiving energy from, a dynamo. Consequently, while it might be possible to use any electric source of energy, such for example as a frictional machine; yet, in practice, the energy is furnished by a dynamo driven by water-power, steam-power, or gas. An ideal form of dynamo exciter would be an *alternating-current* dynamo which generated to-and-fro electric currents, or currents of successively reversing directions, with a frequency precisely that required for setting up resonance in the antenna. If such a very high-frequency dynamo could be constructed conveniently, it would be capable of keeping the antenna in full oscillation indefinitely. That is, if the radiating power of the antenna were say 300 kilowatts or 400 horse-power, it would be possible to connect a dynamo of at least 300 kilowatts capacity (400 H P) to the antenna, and keep it constantly in action at that rate. Such a dynamo would have, however, to generate alternating currents with a frequency either of millions, or, at least, many thousands of cycles per second; whereas the dynamos used in electric lighting and power transmission ordinarily only generate alternating currents with a frequency of sixty

FIG. 31.—Induction Coil for Generating a High Voltage.

(60) cycles per second. This frequency is at least hundreds of times too low for direct excitation.

Under present conditions it is customary to charge the antenna by an induction coil of some kind. When the energy is supplied by a storage battery, an induction coil is used resembling that shown in Fig. 31. This apparatus, which is essentially a powerful spark-coil, has a central core of iron, in the form of a bundle of iron wires. There are two coils, or windings, of insulated wire placed on the iron core. These two windings are carefully insulated from the core, and from each other. One is the primary winding, consisting of comparatively few turns of coarse cotton-covered copper wire. The other is the secondary winding of very many turns of fine silk-covered wire. The primary wires are led out at *p p* and the ends of the long fine secondary winding are connected to the discharge knobs *s s*. When a strong current is flowing steadily through the primary winding, supplied by an external storage battery, there will be no electric impulse, or *electromotive force*, in the secondary. There will, however, be a powerful stationary magnetic flux distribution surrounding the primary current, and linked with the secondary coil. If the primary current be now suddenly interrupted, the magnetic flux linked with the coils will collapse and disappear. In so doing, however, its movement generates a brief but very powerful electric impulse in the secondary winding, constituting a powerful electromotive force, or a high *voltage*, *i.e.*, a voltage capable of jumping across a considerable distance of air-space. Other things being equal, the length of the air-gap across which a spark will jump is an indication of the magnitude of the electromotive force or voltage producing the spark.

### Similarity of Process of Transferring Energy in Induction Coil to Wireless Transmission

The conditions which accompany the transmission of electric power from the primary to the secondary winding, a distance of a few millimeters or centimeters (a few tenths of an inch up to an inch or two), resemble those which accompany the transference of electric energy from the sending to the receiving antenna. Whereas, however, in the latter case, the distance between the primary and secondary wires is relatively very great, and the energy is transferred from one place to the other stowed away in a wave or series of waves; in the former case of the induction coil, the wave has no room to develop a separate existence, but the electromagnetic fluxes are linked with both circuits throughout the process. For the same reason, the efficiency of the transmission is enormously greater in the induction coil than in the wireless case. Nearly all of the electric energy leaving the primary winding is absorbed by the secondary winding. On the contrary, nearly all of the electric energy leaving the primary antenna goes off into space, or else is ultimately absorbed in the ground, and hardly any is absorbed by the secondary antenna. In the rough calculation given on page 95, Chapter VIII, it appeared, for example, that only 1 part in 5,655,000 of the energy liberated by the oscillator, or sending antenna, was picked up by the receiving antenna, under the conditions there considered.

### Elements of Sending Apparatus for Producing Electromagnetic Waves

The elements of the connections at a wireless-telegraph sending station are illustrated in Fig. 32. AA is the antenna, or the wire connecting therewith. C is the induction coil. The primary circuit is marked in full lines and the secondary in broken lines. The primary circuit comprises the primary winding of the coil C, the voltaic battery B, a hand key K, and an electromagnetic vibrator or interrupter V. The vibrator may be a separate piece of apparatus included in the primary circuit; or it may form part of the induction-coil mechanism as shown. It is essentially a vibrating circuit-maker-and-breaker like the vibrator of the ordinary electric bell. Its purpose is to interrupt the primary circuit automatically and rhythmically, as long as the key K is depressed. The vibrator V may give interruptions at the rate of say 200 cycles per second.

It also gives a musical note or tone in the surrounding air, corresponding to its frequency of vibration. At each interruption of the primary circuit at the vibrator V, there is a sudden electric impulse generated in the secondary circuit,

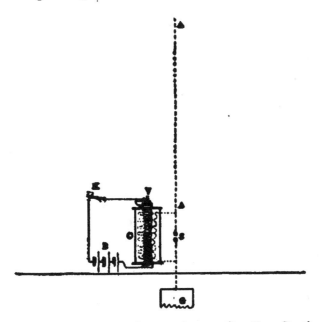

FIG. 32.—Elements of Connections at Sending Station.

and this travels up the antenna at light-speed. If the spark gap g did not break down, there would be a reflection of the impulse from the top of the antenna, accompanied by an electromagnetic impulse or radiation into space; but there would be no succession of waves and no proper development of electromagnetic wave emission. If, however, the impulse on its return from the top of the antenna is able to break down the air-gap g in a spark discharge, the electric oscillation continues, and will go on in a succession of sparks, each feebler than its predecessor and each accompanying a half-wave of radiated energy thrown off into space. After a certain number

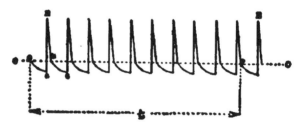

Fig. 33.—Diagram of Electric Impulses Delivered to Antenna by Induction Coil.

of sparks have passed, depending upon the length of the gap and other conditions, the impulse remaining is no longer able to follow up by another spark and the train of dwindling oscillations ceases. Fig. 33 shows diagrammatically the succession of electric impulses generated in the secondary coil under the action of the vibrator V, or at the rate which we have assumed to be 200 per second. The electromotive force rises to E with a sudden jump at each interruption of the vibrator, and then changes more slowly to a

smaller magnitude in the opposite direction.
The sudden kicks *a* E are those which excite oscillations.   Ten of these kicks are indicated in
one-twentieth of a second.   Each impulse occurs
in one two-hundredth of a second, which is
$\frac{5,000}{1,000,000}$ of a second or 5,000 microseconds.

### *Discontinuity of Electromagnetic Flashes from a Coil-Fed Antenna*

Fig. 34 indicates the surges or oscillations set
up in the antenna at every kick or impulse a E

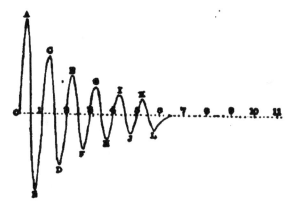

FIG. 34.—Diagram of Oscillations Set Up in the Antenna
at Each Electric Impulse in Secondary Coil.

in Fig. 33, as set up by the vibrator V of Fig. 32.
The illustrations show six complete waves A, C,
E, G, I, K or twelve successively reversed im-
pulses of steadily diminishing amplitude.   The

last impulse L is supposed to be too feeble to
create another spark at the gap g of Fig. 32, so
that the series of oscillations comes to an end
at L. Each complete or double oscillation is
represented as occupying one microsecond, corresponding to an emitted wave-length of 300
meters (328 yards) or a height of simple rod
oscillator equal to 75 meters (82 yards).   It is
evident that the whole series of six complete
oscillations only lasts for 6 microseconds, and
since the kicks, or stimuli, from the induction
coil, only occur by assumption at intervals of
5,000 microseconds, there is evidently a long
interval of darkness and inactivity between the
little flashing intervals in which the antenna is
giving out waves of invisible light, or long-wave
polarized light.   The period of darkness is in this
instance 832 times as long as the period of light.
If the key K in Fig. 32 were held steadily down,
and the vibrator V were thus kept at work, a distant eye assumed capable of seeing this long-
wave light would see the antenna shine out, in a
certain unknown color, for 6 microseconds in
every 5,000, like a flashing lighthouse which sent
a beam over the sea for 6 seconds every eighty-
three minutes.   Although the numerical values
here assumed may vary in practice through a
considerable range, yet they are fairly represen-

tative, and a wireless telegraph sending antenna in full activity is many times more intermittent than the longest period flashing light house in the world.

The observer with the hypothetical eye capable of perceiving the long electromagnetic waves of wireless telegraphy would always see the flashes in the direction of, or "on the true bearing," as a sailor would say, of the sending antenna, but the flash would appear tangential to the surface of the ocean, or in the true level horizontal plane, as distinguished from the actual visible horizon which is depressed somewhat below; so that the observer would expect to see the luminous image of the antenna thrown up, as though by mirage, to the level of his eye or to the horizontal plane at his level. Long after the antenna ceased to be visible by ordinary short-wave light, which moves in radial straight lines, he would expect to see the flash of the antenna by the bending of the long waves around the conducting curved surface of the sea.

### Nature and Use of Auxiliary Condenser at Sending Antenna

An important piece of apparatus auxiliary to the antenna when set in oscillation is a "condenser." It consists of an expanded pair of opposed conducting surfaces, such as tin foil, separated by relatively thin sheets or intervals of insulating material, such as glass, mica, oiled paper, oil or compressed air. A simple condenser may be formed of a sheet of window glass, coated on each side with tin foil, except near the edges. The thinner the glass and the larger its surface area, the greater the electric charge it will hold, or the more electric flux and electric flux energy it will stow away in the glass, for a given magnitude of charging voltage, or in technical language, the greater becomes its *capacity*. Another well-known form of condenser is a glass bottle, coated on the inside, as well as on the outside, with tin foil. It is the glass walls of this bottle or Leyden jar, which receive the electric flux and flux energy when the jar is charged. The greater the surface of the jar and the thinner its wall, the greater will be its capacity.

Looking at an antenna as a condenser, or Leyden jar, the surface area of the conductor or conductors composing it may be considerable; but the slab of insulating air between the antenna and the ground is on the average many meters thick. Consequently, the capacity of an antenna is relatively small. An antenna 50 meters (54.7 yards) high, even if made up of numerous wires, may have no more capacity than a single Leyden jar of ordinary size.

A diagrammatic view of the electric flux stowed away in the dielectric of a charged condenser is shown in Fig. 35, where the upper conducting plate is represented as being charged positively. The flux density increases with the thinness of the insulating slab and also with the charging voltage. A limit to the thinness of the insulator is set, however, by the electric strength of the material, which ruptures, or breaks down in

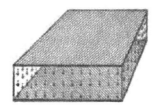

Fig. 35. — Diagrammatic Section of a Charged Condenser Formed by Two Parallel Plates, Showing the Distribution of Electric Flux in the Insulator Between Them.

spark discharge, if a certain electric intensity is exceeded. Air at an ordinary pressure and temperature has a strength (between parallel planes) of about 4 kilovolts (4,000 volts) per millimeter (101,600 volts per inch), glass 8 kilovolts per millimeter, mica 25 kilovolts per millimeter, and so on for other substances. The electric strengths are affected by the purity of the material, its temperature and other conditions.

### Adjustment of Auxiliary Condenser Circuit to Consonance with Antenna

A condenser is often connected in parallel with the antenna at the sending station in the manner

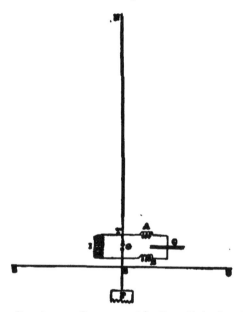

FIG. 36.—Condenser Connected in Parallel with Antenna to Re-enforce Oscillations.

indicated in Fig. 36. By this means the capacity of the insulated system may be much increased so that it will receive a much greater electric charge from the induction coil, with correspondingly increased electric flux and electric flux energy. When the system of Fig. 36 is discharged at the spark-gap, the energy released in radiation may be considerably increased, owing to the presence of the condenser and its electric flux contents. On the other hand, however, the auxiliary circuit containing the condenser should be adjusted to the length of the antenna, in such a manner that the two shall oscillate together, or in synchronism. In other words, the time of oscillation of the condenser circuit A C B G should be adjusted to the time of oscillation of the antenna; otherwise, the oscillations of the two will mutually interfere, and cancel each other; so that if the condenser circuit is not tuned into synchronism with the antenna, the radiation into space may be weakened, instead of being enhanced, by the presence of the condenser, in spite of the greater stock of energy available.

The tuning of the condenser circuit may be accomplished not only by altering the size of the condenser, but also by altering the length and disposition of the wire connecting the condenser to the induction coil. If this wire be arranged in a coil A or B, Fig. 36, of several turns, the effect of the turns is to increase the virtual length of the wire in rapid proportion, because the magnetic flux generated around any one turn

links also, more or less, with the other turns.
By suitably adjusting the number of turns of
wire in the condenser circuit, the free period of
oscillation discharge of the condenser can be adjusted into synchronism with that of the antenna;
so that the latter can thereby be thrown into re-
enforced oscillation and wave emission.

### Loaded Antennas

It is also possible to alter the virtual length of
the antenna, by connecting a coil of a few turns
of wire in its circuit as indicated in Fig. 37. Such
an antenna is called a *loaded* antenna to distinguish it from the simple or unloaded antenna of
Fig. 26. An antenna loaded by a simple coil as
in Fig. 37 always behaves as though its length
were increased. That is its wave-length is increased, or its frequency of oscillation is reduced.
Whereas, therefore, a simple antenna, say 25
meters (27.3 yards) in height, would throw off
waves 100 meters (109.4 yards) long, or with a
frequency of 3,000,000 cycles per second, or with
a period of $\frac{1}{3}$ of a microsecond; the same antenna
loaded with a coil might readily increase its wave-
length to a kilometer or more (1094 yards) with
a frequency of 300,000 cycles per second. At
the same time, however, the radiating power of

the antenna, or the energy it can throw off
in a single wave is likely to be greatly reduced
by loading.

It is thus possible to adjust the antenna and
the condenser into synchronism by altering the

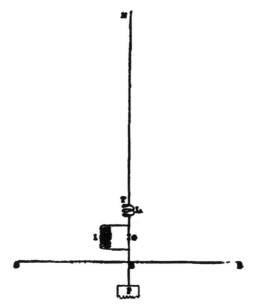

FIG. 37.—Coil Inserted in Series with Antenna to
Increase Its Virtual Length.

number of turns of wire in circuit with either.
It is also possible to lengthen the wave emitted
by an antenna, within certain limits, by loading
it with an appropriate coil, releasing the energy
in a longer train of feebler waves instead of a
very short train of stronger waves.

The length of the wave commonly employed in ordinary wireless telegraphy varies from say 100 meters to 10 kilometers (109 yards to 6.2 miles) corresponding to frequencies between 3,000,000 and 30,000. A common wave-length would be, say, 300 meters (328 yards) with a frequency of 1,000,000 cycles per second.

It may be observed that the presence of the induction coil I in Figs. 32, 36 or 37 does not appreciably affect the virtual length of the antenna, because it is in parallel to the spark-gap G and not in series therewith, like the coil L of Fig. 37. If the induction coil I were thrown in series, thereby adding its virtual length to the antenna, the frequency would be insignificantly low. On account of the very large number of turns in the secondary coil I, or, as it is technically expressed, on account of its large *self-induction*, the oscillations set up on the antenna pass across the gap G and are unable to find their way through the wire of the coil.

It is also possible to insert a condenser into the path of the vertical antenna with the tendency of increasing the frequency of oscillation, or of diminishing the virtual height of the antenna. The actions of a condenser and a coil are in this respect opposite to each other.

*Danger of Secondary Internal Reflections Occurring in a Loaded Antenna.*

When, however, any sudden obstacle or apparatus, such as a coil, a condenser, a resistance, or a discontinuity of any kind is inserted in the path of an antenna, there is a tendency to set up reflections of the oscillations at the discontinuity. These reflections may break up the rhythm and diminish the amplitude of oscillation. Consequently, care has to be taken so to introduce discontinuities into an antenna as to minimize the detrimental effect of, internal partial reflection; or the benefit gained by the insertion of the discontinuity, as in adjusting the frequency to resonance, may be more than offset by the shattering of main oscillations into minor and discordant ripple trains.

# CHAPTER X

## *Voltage Detectors and Current Detectors*

SINCE the human eye is incapable of responding to the long-wave flashes, or electromagnetic waves, given off by a wireless telegraph antenna, an artificially constructed eye has to be used in order to detect and respond to them. The plan followed is to place an antenna at any suitable place in the path of the waves, so that oscillating electric impulses may be set up in this antenna, and then to permit these impulses to act upon some electromagnetic apparatus connected either directly in the path of the antenna, or indirectly, by the aid of a little induction coil. The receiving instrument must therefore be affected either by the oscillatory voltage, or by the oscillatory current in the antenna.

A voltage detector may be theoretically any apparatus which responds to electric potential difference; such as an electroscope or a pair of diverging gold-leaves. In practice, however, it consists of a little instrument called a *coherer*.

A current detector may be of any of the various types which are used to indicate the presence of high-frequency alternating currents. There are a number of different receivers of which we need only consider the prominent types. In practice, there are three well-known types: namely, the *thermal*, the *electrolytic* and the *electromagnetic*.

### Coherers

Coherers are illustrated typically in Figs. 38, 39 and 40. In Fig. 38, we have a sealed glass tube T T about 4 cms. ($1\frac{1}{2}$ inches) long and of $2\frac{1}{2}$ mm. ($\frac{1}{10}$ inch) bore. Near the middle of the tube are two metallic plugs, P P, often made of silver. These are connected to the external wires WW by sealed-in platinum connections. The plugs P P do not touch, but are separated by a small gap about $\frac{1}{2}$ mm. ($\frac{1}{50}$th inch) wide. The tube may be partially exhausted of air; but the gap between the plugs P P contains fine metallic powder, or metallic dust, which lies loosely in the little crevasse.

The loose metallic particles bridging across between the plugs P P have the property of offering an obstruction, or very high resistance, to the flow of current from a single voltaic cell. In other words, the gap is almost an insulator to this feeble voltaic electromotive force. If, however, a higher electromotive force be applied across the gap for even a very minute interval of time, its effect is to break down the insulator and

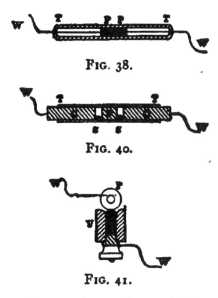

FIG. 38.

FIG. 40.

FIG. 41.

FIGS. 38, 40 and 41.—Types of Coherers.

to allow the voltaic cell to send a continuous current. The sudden higher electric impulse changes the resistance offered by the bridge of metallic dust from a very high to a relatively low value.

The exact nature of the action which takes place when the electric impulse operates, and when the resistance of the gap breaks down, is hard to determine with certainty. It has been much discussed and unanimity has not yet been reached upon the matter. It suffices for present purposes, however, to say that the extra voltage brought to bear across the gap of metallic particles causes them to weld together, or to *cohere* electrically, thus converting a very bad joint in the local circuit of the voltaic cell into a fairly good one.

### Connections Between Coherer and Antenna

The simplest method of connecting the coherer of Fig. 38 with the receiving antenna is indicated in Fig. 39. A B C S G is the antenna path to ground. It is cut at C, and the coherer inserted by means of the wires W W, Fig. 38. A *local circuit* E M C, indicated in broken lines, connects a suitable low voltaic electromotive force, such as a single voltaic cell, to the coherer terminals through an electromagnetic receiver, represented as an ordinary wire-telegraph sounder. Prior to the arrival of electromagnetic waves, the gap of filings in the coherer interposes a high resistance in the local circuit, as well as in the antenna path. Consequently, no appreciable current flows through the sounder M, the armature lever of which remains released against its

upper stop under the action of a spiral spring.
As soon as an electromagnetic wave, or wave-
train, of suitable intensity passes the antenna, an
oscillating electromotive force will be set up along
the antenna and across the coherer gap. The

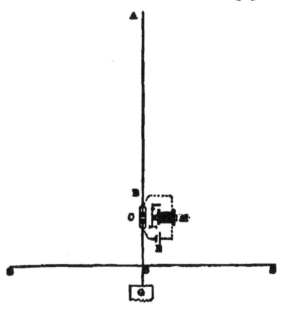

Fig. 39.—Essential Elements of Coherer Connections
When Receiving Signals.

coherer will instantly break down in insulation,
and will cause the metallic filings to cohere. The
oscillations will not discharge through the local
circuit owing to the self-induction or choking
effect of the coil on the magnet M. The voltaic
cell E will now be able to send a current through

the local circuit E M C and excite the electro-magnet M of the sounder, the armature lever of which will descend with a click, thus giving evidence of the arrival of the wave.

## Mechanical Decoherence

The current in the local circuit would continue indefinitely after the bridging of the coherer gap by the first wave, if means were not provided for *decohering*, or restoring the coherer to its original insulating state. This may be done by giving a tap or light mechanical agitation to the coherer tube, thus shaking up the filings in the gap and breaking up the recently welded bridge between the plugs P P, Fig. 38. In practice, the armature lever of the sounder M, may be arranged to deliver a light tap to the coherer tube at the same moment that it produces its click. This tap restores the original condition of the coherer, interrupts the local circuit and cuts off the excitation from the sounder magnet M, which promptly releases its armature lever under the influence of the spiral spring; so that the apparatus is again ready to respond to the next electromagnetic wave.

The connections of the local circuit are usually somewhat more complex than Fig. 39 shows; but the principle remains essentially the same.

The sounder M, for instance, is not directly actuated by the local circuit of the coherer, because it needs a relatively strong current, which is unsuitable. A delicate electromagnet called a *relay* (see Figs. 60 and 61) is, however, placed in the circuit at the point M, and the armature lever of the relay closes another local circuit through a more powerful voltaic battery and the sounder. A feeble current through the local circuit of the coherer is thus enabled to send a suitably strong current through the sounder. An auxiliary electromagnet, actuated also by the relay, is often applied to the sole duty of tapping the coherer, or causing it to decohere, after the relay and sounder have responded.

It is evident that when the apparatus is in working order, signals consisting of short and long groups of electromagnetic waves will be able to spell out corresponding short and long operations of the electromagnet M, or dots and dashes of the Morse alphabet.

Fig. 40 represents a modified form of coherer in which there are two gaps g g. In each gap there is a little globule of mercury. The end plugs C C may be of carbon, and the central plug a little cylinder of iron. This form of coherer has the advantage that it is self-decohering, or auto-decohering. That is to say it needs no blow or agitation to restore the *status quo* after the passage of a wave. It normally possesses a high resistance. The electric oscillation or surge in the antenna breaks down this resistance momentarily and permits a current to flow through a local voltaic circuit. Immediately after the passage of the wave the high internal resistance is restored. The reason for this remarkable action is concealed in the general obscurity of the whole subject of coherence, but is perhaps connected with the liquid state of the substance in the gaps.

Another form of coherer is indicated in Fig. 41. It consists of a small insulating vessel or reservoir V, containing mercury. The mercury is brought into very light contact with the thin edge of a metallic disk P, kept rotating by clockwork. There is a very thin film of insulating oil on the surface of the mercury and the effect of the film is to insulate, or electrically separate, the metallic disk from the mercury. The thin film, may, however, be broken down, or electrically disrupted, by a relatively feeble voltage in excess of that used in a local voltaic circuit. The wires W W connect the device with the local circuit and the antenna, as in Fig. 39. On the arrival of a signal, the oil film is broken and contact established between the disk and mercury; but the revolution of the disk almost instantly re-stores the oil film and decoheres the device.

### Hot-wire Receivers

Coherers depend, as we have seen, upon the electromotive force or voltage of the surge set up in the antenna to force a discharge across the gap

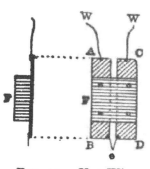

FIG. 42.—Hot-Wire Receiver.

of imperfectly contacting matter, in order to give passage to a local voltaic current. Among receivers which depend, however, upon the magnitude of oscillating current set up through them when they are inserted in the path of the receiving antenna, without interrupting the same, we have the hot-wire receiver. One form of this device is presented to view in Fig. 42. A pair of parallel brass strips AB, CD, are fastened near to each other and side by side, by an insulating block F of hard rubber. Leading-in wires W W are soldered to these strips above, at A and C. Between the lower adjacent corners is soldered a little piece of silver wire e, bent into the form of a sharp V. This silver wire may be about 3 millimeters (0.12 inch) long and about

0.076 millimeter (0.003 inch) in diameter. A cross section of this silver wire is indicated in Fig. 43, at A B C. At or near the center q is a thin filamentary wire of platinum, like the wick inside a paraffin candle. The diameter of the platinum wick is about 1.5 microns (0.0015 millimeter or 0.000,06 inch), or about one-fortieth of the diameter of a thin human hair. A platinum wire so fine is only obtained by thickly coating an ordinary size of platinum wire with silver, and then drawing down the thick composite wire

FIG. 43.—Cross-Section of Composite Wire.

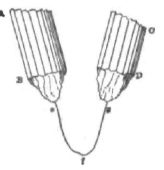

FIG. 44.—View of Loop of Sensitive Fine Wire Under Microscope.

through successively diminishing dies. As the silver wire gets thinner and longer, so also does the internally held wick or filament of platinum. After the little V loop of silver candle-wire has been soldered to the brass plates at B and D, Fig. 42, the device is carefully lowered into a bath of nitric acid, in such a manner that the point of the V loop is submerged in the acid, which immediately attacks and dissolves the silver chemically, leaving the platinum wick un-injured. The process is aided by a feeble electric current from a local voltaic cell, is watched under the microscope, and is arrested at the proper stage. The appearance in the microscope of the V loop after the silver has been dissolved off the tip is shown in Fig. 44, where A B and C D are the 76-micron or 0.076 millimeter silver wires, and e f g the 1.5 micron platinum filament, hanging in a short loop. The device is then ready for use and is conveniently protected from injury by placing it in a short glass bottle or test-tube.

The connection of the little hot-wire device with the receiving antenna is illustrated in its simplest elements at Fig. 45. A B is the antenna, connected to ground through the hot-wire at H. A local voltaic circuit, in broken lines, connects a feeble electromotive force, such as a single voltaic cell, through the telephone receiver T and the hot-wire H. Prior to the advent of electromagnetic waves, a steady current flows through the local voltaic circuit, producing no sound in the receiver T. This current serves to warm the fine platinum wire, the electric resistance of which is appreciable, but constant at any constant temperature. As the temperature of the platinum is increased, however, the resistance increases.

If now an electromagnetic wave or wave-train strikes the antenna **B A**, an oscillating current will pass through the fine platinum filament **H**, and will heat the same appreciably, being superposed upon the steady current from the voltaic

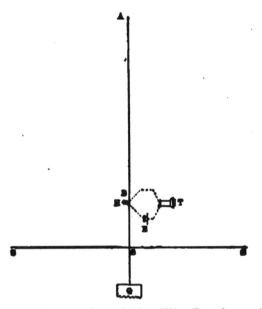

FIG. 45.—Connection of Hot-Wire Receiver with Receiving Antenna.

cell **E**. The antenna is prevented from discharging to ground through the telephone **T**; or by the path **AB TE SG**, owing to the self-induction, or choking effect, of the telephone. Practically all the discharge goes through **H**. The momentary increase in the heat and tem-

perature of the filament H causes its resistance
to be momentarily raised, and this reacts upon
the local voltaic current, diminishing the same
momentarily.    The telephone T responds audi-
bly to the sudden alteration of current, which
lasts as long as the waves or groups of waves are
passing, and ceases the moment the waves cease
to arrive.    The sensitiveness of the device is due
to the small dimensions of the fine filament.    The
oscillating electric currents received through the
filament from the antenna are very feeble when
the antenna is far from the sending station; but
the cross-section of the filament being only about
2 square microns ($\frac{1}{800}$ of one millionth of a
square inch), even a very feeble electric current
will be condensed to a relatively appreciable current density at the filament, thus giving rise to
appreciable heating in a mass of metal only
about 2,000 cubic microns in volume ($\frac{1}{1,600}$th of
one millionth of a cubic inch).

## Electrolytic Receivers

The device, represented in outline by Fig. 46,
consists of a small vessel C containing a suitable
solution, such as dilute nitric acid.    A candle
wire w, of the kind above described in connection
with Figs. 43 and 44; i.e., a silver wire of about
76 microns in diameter (0.003 inch) with a

platinum wick about 1.5 microns in diameter (0.000,06 inch), is immersed to a suitable depth, perhaps a quarter millimeter (0.01 inch) in the solution. The acid dissolves off the silver, so that the filament of plat-inum is immersed in the solution, offering thereto an immersed surface area of about 1,200 square microns ($\frac{1}{2,000,000}$th of a square inch). The wire w is fastened to the lower end of a brass screw hav-ing a milled head ss, the screw passing through a brass support P P. The depth of immersion of the

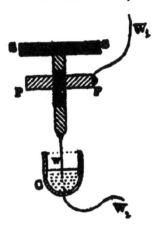

FIG. 46.—Electrolytic Receiver.

fine platinum filament can be adjusted by turn-ing the milled head in one or the other direction.

A solution of an acid or alkali traversed by an electric current is called an *electrolyte,* and the current is carried only by atoms, or groups of atoms, which are separated out from the solu-tion. In other words electric conduction through an electrolyte is accompanied by chemical de-composition of the electrolyte.

The connection of the electrolytic receiver with the receiving antenna is essentially the same as

that of the hot-wire receiver represented in Fig.
45. That is, a local voltaic current is provided
containing a small electromotive force and a
telephone receiver. Prior to the advent of the
electromagnetic waves, the cell E, Fig. 45, sends
a feeble steady current through the telephone
and the electrolytic receiver. This current
causes minute bubbles of gas to be liberated from
the fine immersed platinum filament, but the
telephone T gives no sound. As soon as an
electromagnetic wave strikes the antenna, the
oscillating current set up passes through the
electrolytic receiver, and heats the same in the
minute constricted mass of liquid immediately
surrounding the fine platinum filament. The
effect of the heat so liberated is two-fold. In the
first place it momentarily raises the temperature
of the pellicle of solution immediately surrounding the filament, thereby reducing the electric
resistance of the device; for electrolytes, unlike
metals, improve in conductivity when heated. In
the second place the gas is liberated more freely
from the surface of the fine filament as the temperature is increased. In technical language,
the momentary warming effect of the oscillating
current causes the filament to be partly *depolar-
ized*. Owing to both of these actions, the appar-
ent resistance offered to the local current from

the voltaic cell is temporarily diminished and the sudden increase of current produces a sound in the telephone. Immediately after the passage of the wave, the heat is dissipated by conduction into the solution, and the original resistance and *counter electromotive force of polarization* in the device are restored.

The sensitiveness of this device is attributable to the constriction of the conducting path from antenna to ground into a minute volume of liquid, having an extremely small cross-section. In this tiny volume there exists a very appreciable electric resistance and also an appreciable electrolytic back voltage, like that of an opposing voltaic cell. The thermal effects of even a very feeble oscillating current from the antenna are here condensed into so small a volume that the temperature of that small volume can be appreciably raised. The rise of temperature has a powerful effect both on the localized resistance of the constricted liquid path and on the back voltage of the virtual opposing voltaic cell contained in the device.

Another form of the electrolytic receiver is represented in Fig. 47. A vessel, V V V, such as a small glass tumbler, is filled with an electrolyte such as dilute nitric acid, or dilute caustic soda, to the level L L. Two metallic surfaces or *electrodes* dip into the electrolyte. One of these, indicated at E, may have any convenient size and shape, connecting with a leading-in wire W,; or this wire may itself dip into the solution and form the electrode E. The other

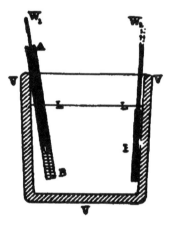

FIG. 47.—Simple Form of Electrolytic Receiver.

electrode like *w*, in Fig. 46, has extremely small surface area. A very simple way of preparing such a small area of electrode is illustrated in greater detail at the lower part of the Figure. A glass tube, a b c of any convenient dimensions, say 7.5 cms. (3 inches) long, 3 mm. and 1 mm. (about $\frac{1}{8}$ and $\frac{1}{25}$ inch) in external and internal

diameters respectively, is slipped over a short
length of copper wire. This copper wire is
welded at one end in the flame of a Bunsen
burner to a few centimeters (one inch, say) of
fine platinum wire having a diameter of about
0.05 mm. (0.002 inch). In the illustration, the
weld or junction between the copper and platinum wires is shown at J. The tube is then
heated to the softening point over the fine platinum wire, and the softened walls are squeezed
tightly over the platinum wire with a small pair
of tongs, so as to seal in the wire hermetically for
a short distance, say 1 cm. ($\frac{3}{8}$ inch). The tube
after cooling is now scratched with a file across
the seal and is broken sharply across with the
fingers. The break if properly made will leave
the ruptured platinum wire very slightly projecting beyond the glass. A few strokes, with a fine
file, will file the platinum wire flush with the
glass, thus presenting as at $d$, Fig. 47, an exposed
disk of platinum of the diameter of the fine wire
at the end of the seal b c, and with a surface area
of about 2,000 square microns or $\frac{3}{1,000,000}$ square
inch.

The depth of immersion of the glass tube in
the electrolyte is of no consequence so long as the
end of the fine platinum wire is fairly covered by
the solution. Neither does the distance between

the two electrodes in the vessel V V, make any appreciable difference. In other words, the seat of the actions occurring in the apparatus is in the constricted liquid path immediately covering and including the minute exposed area. The resistance of the device is practically all located within a millimeter ($\frac{1}{25}$ inch) of the exposure, and the voltaic counter electromotive force of polarization is located at the exposure; so that all the rest of the electrolyte merely provides an outward escape for the electric current that has passed through the intensely constricted region of influence over the minute area.

This form of electrolytic receiver is probably the simplest to construct of all the receivers used in wireless telegraphy. Like most of the other devices it is patented. It is not capable of adjustment in surface exposure area, like the apparatus of Fig. 46, and if the minute exposure gets dirty or clogged, it has to be thrown away and a new one substituted. It is, however, capable of great sensitiveness and the materials for its construction are very easily obtained.

### Electromagnetic Receivers

There are various types of electromagnetic receivers, but that illustrated in Fig. 48 is generally admitted to be the most sensitive. It con-

sists of a flexible band b b b b of iron wires passing over the grooved pulleys L L, which are
steadily driven by clockwork. The band of iron
wires moves through a glass tube t t, on which is
placed a winding of insulated wire with external
connections W W, leading to the antenna on one
side and to ground on the other. This winding,

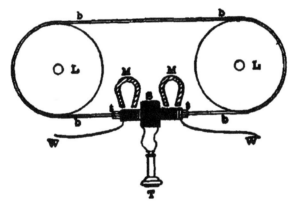

Fig. 48.—Magnetic Receiver.

inserted in the antenna path, forms the primary
winding of a little induction coil, in which the
moving band of iron wires is the core. The
secondary winding S, placed over the middle of
the glass tube is connected to a telephone receiver, T.

The band of iron wires in passing through the
tube makes its procession in front of the two
fixed permanent horseshoe magnets M M. These

are so arranged with regard to strength and
direction of polarity, that the band of iron
emerges from the tube with its internal magnetism reversed in direction from that with which
it enters.

Iron has a curious magnetic property when its
magnetism is cyclically reversed. If the magnetism is established along a band of iron wires
in one direction, then when the process of de-
magnetization and reversal is started, the change
of magnetic flux in the iron takes place very
slowly at first, until a certain stage of magnetic
instability is reached, and then the magnetic flux
reverses with great swiftness. The action may
be compared to that of a ball moving alternately
from side to side on the deck of a rolling ship at
sea. If the deck is flat and plane, the ball will
swing regularly from side to side. If, however,
the deck be somewhat bowed, rising in the middle
like a turtle-back, the ball will be slow to return
on each roll until it gets to the top of the turtle-
back and then it will run down with great speed.

The function of the two magnets M M in the
magnetic receiver is to bring the iron wire core
of the induction coil into the unstable magnetic
condition during its passage within the primary
winding connected in the antenna path. Under
these conditions, if any electric oscillations come

through the primary coil from the antenna, they will be able to shake out the magnetic flux in the enclosed band and precipitate its reversal. The rotating mechanism brings the magnetic flux to the edge of the precipice, as it were, and the feeble electric currents are able to push it over. The sudden change of magnetic flux inside the secondary coil $s$, sets up an electric impulse that will produce an audible sound in the telephone T.

The above form of magnetic receiver is thus essentially an induction coil with the antenna path passing through the primary winding and the delicate receiving telephone in the secondary. The induction is increased in sensitiveness by the aid of the constantly renewed magnetic instability in the iron core, under the action of the permanent magnets. .

### Comparison of Receivers

Comparing the behavior of the various types of receiver, it is to be noted that the coherer is the only one which permits of a permanent record being obtained. The coherer, as outlined in Fig. 39, operates an electromagnetic receiver of the Morse type. Such a receiver is able to record the message in dots and dashes inked upon the surface of a long strip of paper, coiled on a roller in the apparatus and moved by clockwork. A particular form of Morse inkwriter is seen in Fig. 49. It will be observed that the armature consists of a split soft iron tube fastened to a rocking lever in such a manner as to be attracted downward when the black coated electromagnet

FIG. 49.—Morse Inkwriter.

is excited. The lever throws up a disk against a moving band of paper not shown. The disk is kept rotating by clockwork and dips into an inkwell.

On the other hand, however, the speed at which signals can be received and recorded by means of the coherer is distinctly lower than that obtainable with non-recording receivers. A speed of 15 to 20 words a minute is considered good with a recording receiver. With some non-recording receivers this speed may be doubled.

In regard to sensitiveness, the coherer has hitherto proved much inferior to the others. The most sensitive is the electrolytic receiver and next to that the magnetic. Both of these use the telephone as the receiving instrument.

### Telephone Receivers

A convenient form of telephone receiver, illustrated by Fig. 50, is such as telephone operators employ. A leather-covered steel band L L goes over the head and supports the receiver R R close to one ear. The band is fastened to the receiver by the thumb-screw *s*. The covered wires w w serve to connect the receiver with the antenna system.

FIG. 50.—Head Telephone.

Fig. 51 shows the parts of the receiver disassembled. B B is a hard rubber box with a screw cover C. Inside the box are three pairs of half-ring steel permanent magnets, NS, NS. In the center, a pair of soft iron pole pieces are supported, receiving their polarity from the magnets NS and wound with many turns of fine silk-covered copper wire connecting with the leading wires *ww*. D is the ferro-type disk of steel which is clamped around it's edge between the box and the cover, so as to be held over but not quite touching the poles at the center.

The sensitiveness of the electrolytic and magnetic receivers is at least partly attributable to

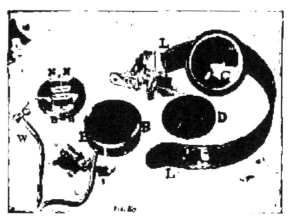

FIG. 51.—Head Telephone Disassembled.

the great sensitiveness of the telephone which they employ as their intermediary with the human brain. The telephone, as is well known, is extraordinarily sensitive in detecting feeble electric currents undergoing rapid variation.

# CHAPTER XI

## Alternate Use of an Antenna for Sending and Receiving

In order to carry on simple wireless telegraphy between a single pair of stations, remote from all other wireless telegraphists, it is evidently necessary to have an antenna at each station. The dimensions required for the antennas will depend upon the distance between them. For sending messages from one room to another in the same building, the antennas may be a few centimeters or inches long. For sending messages between adjacent buildings, or buildings separated only by a few kilometers or miles, the antennas need only be a few meters or yards high. For distances of hundreds of kilometers or miles, large and tall antennas are at present necessary. The object of the sending station in long-distance wireless telegraphy is to throw out as long a train of powerful waves as possible, while that of the receiving station is to employ as sensitive a receiver as possible.

149

One and the same antenna is used at a station
for sending and receiving alternately.    The connection is changed from the sending to the receiving apparatus by a switch as indicated in Fig. 52.

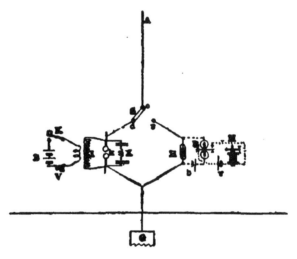

FIG. 52.—Diagram of Switch Connections from Sending
to Receiving.

The switch S, has a metallic blade or lever-arm
which is pivoted at *c* and may be turned into contact with the point *d* for sending, or with the
point *r* for receiving.    A is the antenna, or the
wire leading thereto, and G the ground-connec-
tion.    In the position shown, the switch is turned
to the sending position and the antenna is connected to the spark-gap *k*.    This is excited by
the induction-coil I and vibrator V, when the

key $K$ in the primary circuit is depressed. There is also an auxiliary circuit X, consisting in this case of a pair of condensers with a coil between them, to increase the stock of radiation-energy prior to discharge. The receiving apparatus is indicated as consisting of a coherer H, working into a relay R, (see Fig. 61) through the local circuit H R b. The relay in its turn operates the sounder M through a second local circuit, including the voltaic cell $v$. It is necessary to make sure that the delicate receiving apparatus is completely disconnected from the antenna while the latter is being used for sending. Sometimes an automatic switch is used which will not permit the induction coil to be excited unless the receiver is cut off. In some installations the coherer and its immediate connections are shut up in a metal box to keep accidental waves from impinging upon the coherer.

The sending key used in wireless telegraphy differs only in mechanical details from the ordinary Morse key of wire telegraphy. An ordinary Morse key is illustrated in Fig. 53, and a form of wireless telegraph key in Fig. 54. The wireless key has to send a much stronger current to the induction coil than that which an ordinary wire-telegraph key controls, so its contact is larger and stouter. Moreover, there is a possibility of the operator receiving a severe electric shock from the induction coil; so the insulating handle is made more massive. There is apt to be some sparking at the key contact on breaking

FIG. 53.—Ordinary Wire-Telegraph Morse Key.

FIG. 54.—Form of Wireless Telegraph Morse Key.

circuit, and certain forms of key are designed to overcome this, in some cases by breaking the contact under oil, and in others by breaking the contact between the poles of an electromagnet.

## Morse Alphabets or Codes

In sending signals, the contacts of the key are made by the operator in conformity with the Morse code. There are unfortunately two telegraph codes—the American Morse code, or that in almost universal use on the North American continent, and the international Morse code, or continental Morse code, in use in practically all other parts of the world. The American Morse code was introduced in the early days of the Morse system in the United States. In Europe,

about the same time, a number of different al-
phabets or codes sprang into existence, in different places.   The frequent transition and inter-
communication of messages among European
countries was soon hampered by differences in
Morse alphabet, so that, by international convention, the present Continental code was arrived at and adopted.   Unfortunately, the date
of this convention was prior to the introduction
of Atlantic cables and fast ocean steamers so
that America was not a party to the conference.

The two codes are presented in Fig. 55, so far
as concerns the use of the English language.   It
will be seen that the signals for a, b, d, e, g, h, i,
k, m, n, s, t, u, v, w, and 4 are common to both
codes; while the signals for c, f, j, l, o, p, q, r, x,
y, z, 1, 2, 3, 5, 6, 7, 8, 9, 0, . , ?, and ! are different.
In both codes the dot is the standard element of
length.   A dash has the length of three dots,
and the space separating dots or dashes in a
letter are of dot length; except in the American
letters c, o, and z which are called *spaced* letters,
and in which there is an extra space of two dots'
length.   The American l is a six-dot length dash,
and the American zero is a nine-dot length dash.
The space separating adjacent letters is three
dots long and the space separating words, six
dots long.

The American code is shorter on the average, in the signaling of ordinary English, by about 5 per cent.; that is to say 95 dot elements of American code will be equivalent in the formation of letters to 100 elements of international code, so that the American code is swifter by about this amount. On the other hand, the spaced American letters are a source of possible errors, if the signaling is not in the best condition, and some operators who are practically acquainted with both codes maintain that, owing to the care needed in sending spaced letters, there is no sensible difference in swiftness between the codes. It is certain that a person conversant solely with one of these codes is quite unable to read messages sent in the other. Listening to the other code in such a case is like listening to foreign unknown speech. It is greatly to be regretted that this confusion of telegraphic language exists, and wireless telegraphy has tended to make the confusion more evident. Most of the ships carrying wireless telegraph outfits talk international Morse. Only those like the Fall River Liners, on the American coast, talk American. Some ships can talk in either code.

Each dot contact of the sending key must be accompanied by at least one discharge of the induction coil and, therefore, by at least one

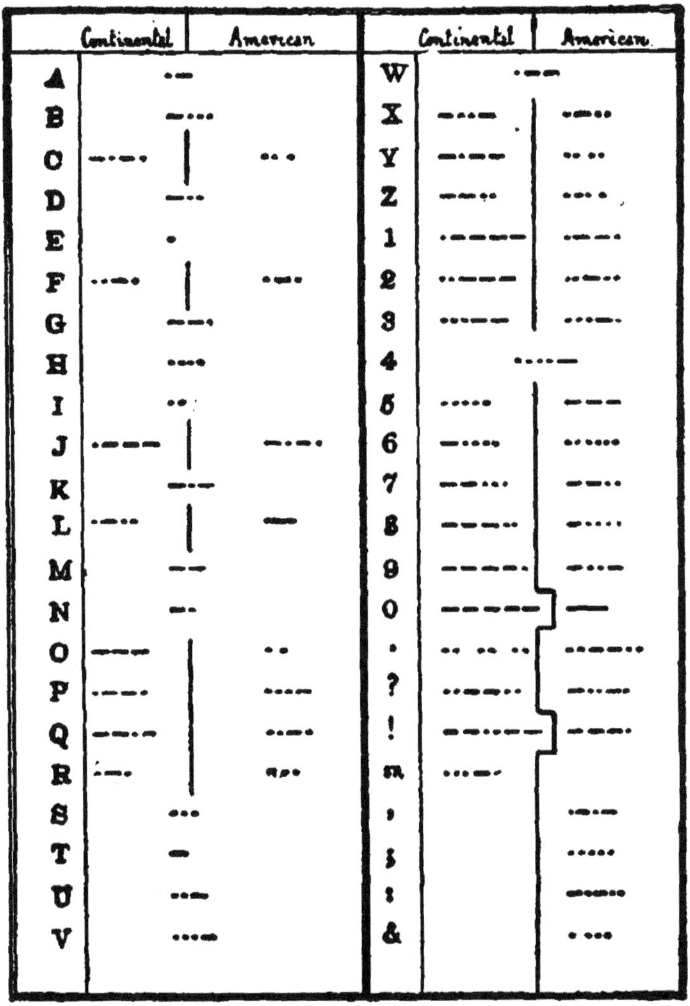

FIG. 55.—Continental and American Morse Telegraph
Codes.

spark, or train of oscillating spark discharges at
the spark-gap *k*, Fig. 52.    The dashes, on the
other hand, need to set up a series of discharges.
A dot may thus be accompanied by a single interruption at the vibrator and a dash by three or
more such interruptions.

### Alternations of Sending and Receiving.

During the time that the message is being sent,
the operator is unable in the ordinary wireless
system to receive a message, or to know whether
the receiving operator has been able to take the
signals.    In ordinary wire telegraphy, as prac-
ticed in the United States, the receiving operator
can "break" the line circuit, and thus notify the
sender that the message is not being taken.    But
the receiver of a wireless telegraph message
cannot ordinarily stop the sender, and the sender
goes on until either the message has been con-
cluded, or until he deems it prudent to turn his
switch and let the receiver send back encourage-
ment to proceed.    For the same reasons, it is not
uncommon for the sender to repeat a message,
as soon as he has finished it, in order that the
distant receiving operator may check what he
has written against the repetition, so as to avoid
mistakes.    Attempts have already been made to
send and receive messages simultaneously by the

same antenna, but such operations can only be regarded at present as in a subsidiary stage. As soon as the difficulties in the path of ordinary, simple, simplex to-and-fro wireless telegraph signaling have been overcome, it will be time to attend to the development of more intricate methods.

## The Insulation of Antennas

The insulator or insulators which support the sending antenna have to be maintained in good order, because they must withstand high voltage without appreciably leaking or sparking. Wherever the antenna wire or wires come in contact with any substance, an insulator has to be used, especially where there is a chance of moisture from rain, dew or fog. The antenna may itself be an insulated or rubber-covered wire. This is, however, of but little service except in preventing atmospheric discharges from driving wind and snow. On the other hand, the insulation of an antenna used solely for receiving messages need not be so carefully maintained. In order to send messages, powerful voltages must be used and insulated; but in order merely to receive messages, or to listen to what is going on in the neighboring ether so far as concerns long electromagnetic waves, only feeble voltages are produced; so that while insulation is proper, it need not be safeguarded elaborately. A bare copper wire fastened around a branch of a tree and touching the boughs or trunk at several places on the way down, may often enable messages to be received. Such a wire could hardly be used for sending messages to a distance.

### Heights of Antennas

Antennas are carried to various heights, as already mentioned. Since the trouble and expense of construction increase rapidly above a height of 30 meters (32.8 yards), very high masts are only installed for very long distance transmission. The greatest height to which they have been thus far carried is 128 meters (420 feet), in the form of a steel chimney. On board ship the mast height usually limits the elevation of the antenna to about 30 meters (98.4 feet). A widely extending antenna is not so useful for receiving as for sending, except when the waves are much longer than four times the mast height; because although side expansion no doubt affords an increase of catchment area for passing waves, there is little doubt that the area from which energy is absorbed by a receiving antenna is fairly wide (see page 84), even when the antenna consists of but a single vertical wire. On the other hand, increasing the height of a receiving antenna increases the energy that can be scooped out of a passing wave-train approximately as the square of the elevation, provided that resonance is maintained; so that increase of height always aids long-distance reception.

The ordinary height of a shore antenna mast is about 45 meters or 150 feet.  If a mast is observed on the seashore with a little wooden house at the base, it is highly probable that it is constructed for wireless telegraph purposes.  If a group of several wires can be seen to festoon from the masthead to the house, the existence, either in the present or past, of a wireless telegraph station may safely be assumed.

Although the usual method of installing an antenna is to have the same insulated at the top and well grounded at the base, yet other methods are possible and are occasionally employed.  For example, the antenna may consist of an arch or vertical loop, either one or both ends of which may be grounded.  Again, the ground connection of an antenna may be dispensed with entirely, and an insulated metallic plate used in its place, the plate being preferably supported parallel to the ground and at a height of about 2 meters (6.5 feet) above it.  Such a plate is equivalent to an air condenser inserted in the antenna path near the ground connection. Or, the ground may be dispensed with, and a horizontal wire may be run at a short distance above the ground. A number of such variations of installation have been suggested or used at different times; but while they introduce differences of action in detail, they usually conform to the same broad, fundamental principles of action as the ordinary lightning-rod type of vertical wire antenna with grounded base.

### Power Required for a Wireless Telegraph Sending Station

In order to receive wireless signals, no power has to be expended beyond the almost infinitesimal amount supplied by voltaic cells in the local circuit or circuits of the receiver. But in order to transmit signals to a distance, very appreciable power must be supplied. For distances under 100 kilometers (60 miles), the power supplied does not usually exceed 3 kilowatts (4 horse-power), and is sometimes considerably less. As the distance increases, the power absorbed is likewise increased. A few stations use 50 kilowatts (67 horse-power) or more. As we have already seen, the antenna radiates the power by jerks, or with long intermissions; so that although in sending with the key depressed, the power supplied to the primary winding of the induction coil may be; say, 2 kilowatts (2⅔ horse-power), the antenna may be varying in power between 200 kilowatts and o.

In the neighborhood of a city, or of an electric lighting system, the power required may be taken from electric-light mains. The same facility may be afforded on vessels electrically lighted. Otherwise, a small gas, oil or steam-engine has to be installed, in order to generate the power required.

### Contrast Between Power Required for Wire and Wireless Telegraphy

The contrast between wire and wireless telegraphy is nowhere more marked than in respect to power. In wire telegraphy, over a distance of, say, 200 kilometers (120 miles), the power required for the transmission of messages is perhaps 8 watts at the generator (5.9 foot-pounds per second, or 0.0107 horse-power). In wireless telegraphy, working over the same distance, 8 kilowatts (10.8 horse-power) may be expended, or a thousand times more. Even then, the wireless system is only rendered practicable by the enormously greater sensitiveness of the receiving apparatus it employs. In the one case, the electromagnetic waves are guided in a single beam to their destination by a wire, and the only serious loss is by leakage over the insulators along the line, or by imperfect conduction and internal warming of the line wire.   In the other case, the power is scattered in every direction and only an extremely small fraction can be picked up and utilized at the distant receiving station.   Nevertheless, the fact that a wire can be dispensed with is so wonderful, that we may be glad to obtain wireless telegraph communication at any reasonable cost, and not be over anxious to cavil at the waste of energy.

### Spreading of Waves from the Sending Station

We have already seen that electromagnetic waves employed in wireless telegraphy start off as hemispherical waves, expanding at light-speed in all directions.   The hemispherical formation fairly commences about a half-wave length from the sending antenna.   Inside that radius the wave is much more complex in form.   Theoretically, the expansion goes on indefinitely, allowing for the curvature of the earth.   Practically, very little has yet been determined about the matter, except that signals from a sending antenna have been detected in balloons at moderate elevations above the earth not far from the antenna, and also by observers on the earth at a distance not exceeding 5,000 kilometers (3,000 miles) in any direction.  No receptions of messages have been yet reported over a greater distance than this.

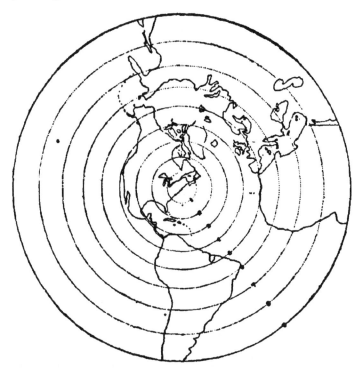

Fig. 56.—The Boston Hemisphere of the Globe in Stereographic Projection; or the Hemisphere with Boston as Pole.

Fig. 56 represents the spreading of the waves from the sending station according to the existing theory.  The sending station is supposed to be located at Boston, Mass., which is, therefore, placed at the pole of the earth, for the purpose of this discussion. Each of the dotted circles 1, 2, 3, . . . 9 represents an over-sea distance of ten degrees, or 600 nautical miles (1,111 kilometers or 690 statute miles). In passing from one circle to the next, the waves will take $\frac{1}{270}$th of a second, or 3,700 microseconds. The complete journey from Boston at the pole, to the equator on circle 9, or one-quarter round the world, would occupy $\frac{1}{80}$th second. It will be observed that the outgoing waves strike Newfoundland, Bermuda and the Great Lakes, about the same instant, $\frac{1}{270}$th of a second after being emitted. Cuba and the Greenland coast would be struck almost simultaneously in $\frac{2}{270}$ths second. When the waves reach England, in about $\frac{4}{270}$ths second, they will also be striking Alaska, the Pacific coast, Norway, France, Spain, Morocco and the Brazilian shore. They will also nearly have reached the North Pole. After $\frac{8}{270}$ths second, Japan, Thibet, Persia, Egypt, the Gold Coast, Argentina and Peru will be reached almost simultaneously. The corresponding development of the waves, in section through the center of the globe and the sending station, is presented in Fig. 57. Boston would occupy the position P, and a point on the ocean south of Perth, Australia, would occupy the antipodes P'.

The expanding waves would occupy the successive shells 111, 222, 333, and 444, the last being
attained in $\frac{1}{18}$th of a second.  The equatorial
circle Q 222 Q corresponds to the circle 9 in

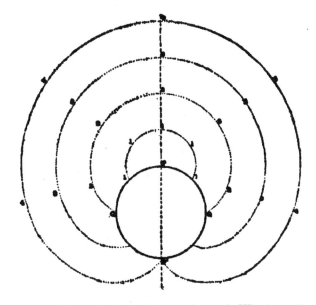

FIG. 57.—Hypothetical Expansion of Wireless Telegraph Waves Over the Globe.

Figure 56.  If will be observed that after the
wave passes the equatorial circle Q Q, it narrows
its circle, and when it reaches the antipodes P',
it has gathered all its feet to this point.  This
should mean that the weakening of the waves,
which occurs by expansion at the outset, diminishes slightly on the surface of the globe after

passing the equator, and the signals at the antipodes P' should be relatively stronger than at
other points in that region. All this is, however,
as yet entirely inferential, because the waves
have not yet been detected beyond the first position 1, 1, about one-eighth of the distance around
the globe, or half way to the equator.

Moreover, it is uncertain as to what happens
to the expanding waves in the upper regions of
the earth's atmosphere, where no balloon can
ascend to take observations. Air is an excellent
insulator at ordinary pressures; but when air is
exhausted from a vacuum-tube, the dregs of air
remaining within can be made to conduct electric
discharges better than sea-water. The pressure
of the air is a maximum at the earth's surface,
and dwindles indefinitely as the earth's surface
is departed from. At an elevation of from 20
to 100 kilometers above the earth (12 to 60 miles),
depending upon local conditions, there must be
strata of rarified air having as low a density as
that produced in such vacuum-tubes. It is uncertain whether very feeble electric forces can
evoke conduction in such rarified air. That is
to say, it is quite possible that conduction may
occur in laboratory vacuum-tubes after the rarified air has been modified, or *ionised* as it is
called—by the action of a relative powerful elec-

tric flux, and that in the absence of such degrees
of electric flux-density, the rarified air would fail
to conduct.  If there is a shell of rarified air
above the earth at a height of, say, 70 kilometers
(roughly 40 miles) which suddenly conducts like
sea-water, then the electric flux would cease to
expand into the space beyond, but would skim
along the interior wall of this conducting shell.
The waves would then be confined to expansion
sideways between the earth's conducting surface
below and the internal wall of the concentric
globe of rarified air above.  This would tend to
reduce the attenuation or weakening of the waves
markedly; because beyond the radius of 70 kilometers, the expansion would continue in but two
dimensions—longways and sideways—instead of
three dimensions—including height.  On the
other hand, however, there might be layers of
rarified air which conducted even under very
feeble electric flux, and yet the conduction might
be gradual instead of sudden.  If the conduction
increased gradually to a maximum through many
kilometers of ascent, there would be loss of energy
by reason of imperfect conduction and eddy-
currents in the transitional layers, so that the
benefit due to ultimate confinement of the waves
within the rarified shells, might be more than lost
by waste of energy in this partial conduction.

The whole subject of wave contour at great distances must remain in abeyance until sufficient measurements have been made of the wave-strengths at different distances to enable the contours to be inferred. Very little information is yet obtainable as to relative wave-strengths at great distances, partly owing to the difficulty of measuring extremely feeble wave-intensities and partly owing to atmospheric variations, to be referred to later. At short distances, *i.e.*, 100 kilometers (about 60 miles) or less, the few results obtained appear to indicate that the energy received diminishes roughly as the inverse square of the distance, or in conformity with simple hemispherical expansion.

*Experimental Apparatus for a Range of a Few Meters*

For simple experimental and demonstrative purposes, apparatus is readily obtainable that will produce recognizable signals at very short range and with an insignificant expenditure of power. The apparatus of Fig. 58 consists of a small induction coil with vibrator, to which the primary current from an external voltaic battery is admitted by the Morse key K. The terminals T t are for connection to a voltaic battery. The secondary winding of the induction-coil charges the insulated double-rod system R R. By this means short waves are thrown off which strike

Fig. 58.—Simple Form of Wireless Sending Apparatus for Transmitting to a Distance of a Few Metres.

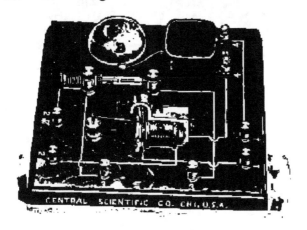

Fig. 59.—Receiver for Experimental Wireless Telegraph Set of Few Metres Range.

the receiver in the vicinity. A form of receiving apparatus capable of being used with such a set is indicated in Fig. 59. Here the horizontal glass-tube coherer C is connected to the binding-posts 3, 3 on the wooden base. Short projecting wires may be clamped in these to help seize the passing waves. A voltaic cell is also connected through binding posts 2, 2 with the coherer, through the relay R. This relay is actuated when the coherer responds, and closes a local circuit through the vibrating electric bell B and

FIG. 60.—Simple Neutral Relay.

another voltaic cell connected to binding-posts 1, 1. The ringing of the bell not only gives the signal, but also agitates the coherer and restores its normal insulation, after the passage of the wave.

The relay R is represented in greater detail in Fig. 60. A pair of electromagnetic coils M are wound with silk-covered fine copper wire connected at the ends to the main terminals T T. The armature of the relay is free to move about a horizontal axis through a small play, set by the two uppermost opposing screws.   A spiral spring adjusted in tension by the set-screw S, draws the armature away from the electromagnetic poles, when these are unexcited by electric current in the coils.   The armature lever then rests against an insulating stop.   On being attracted by the magnet poles, the lever strikes the contact point

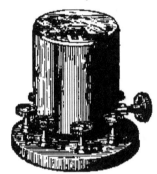

FIG. 61.—Polarised Relay.

C, thus completing a local circuit through the terminals L L, and wires to the same underneath the base.

In long-distance wireless telegraphy with the coherer, a more delicate form of relay is generally employed, such as that shown in Fig. 61. In this form the delicate lever plays, under glass cover, about a vertical axis and the adjustment is provided by the screw on the side.

The simple apparatus of Figs. 58 and 59 for sending messages across a hall, or from one building to another near by, is of some practical importance, because a wireless telegraph station equipped to receive messages over a distance of hundreds of kilometers or miles may not disdain to employ an even still simpler apparatus of the same character, for testing purposes. An ordinary vibrating electric bell, such as that shown at B in Fig. 59, when excited by a voltaic cell, rapidly makes and breaks its circuit at the contact point of the vibrator. Each such interruption is usually accompanied by a little spark at the vibrator contact, and a feeble electric wave, or very brief wave-train is thrown off from the circuit. Such an apparatus may be installed, . say, on the wall of the wireless telegraph station, and excited by pressing an ordinary push button at the receiving operator's desk. The operator wishing to ascertain whether his receiving apparatus is in order, may do so in the absence of any incoming signals, by pressing the bell-button. The feeble electromagnetic waves thrown off from the bell wires are thus enabled to attach themselves to the antenna wire or wires, and so to produce a feeble signal that the operator can recognize.

# CHAPTER XII

## *The Problem of Selective Signaling*

In the last chapter we considered wireless telegraph working between two stations to the exclusion of all others within the working range. But the earth's atmosphere is no longer an Eden with but a single pair of occupants. In most parts of the civilized world, the actual problem is how to communicate with the station that is wanted, and yet to keep out of communication with disinterested stations.

## *Nature of Interference*

When a ship carrying an ordinary untuned coherer receiving set occupies a position at which wave signals are passing from only one sending station, the coherer is able to transmit those signals correctly to the Morse sounder, or inkwriter, in its local circuit. The same may be true if there are two sending stations working simultaneously, one of which produces much

more powerful wave signals, at the ship's position, than the other. The adjustment of the
coherer may be such that the signals which are
feebler, either owing to greater distance, or
weaker transmitting apparatus, are unable to
interfere, and only the stronger signals are recorded on board the ship. If, however, there
are two or more stations in the neighborhood
sending signals simultaneously, and the intersecting waves from these stations are about
equally strong, the ship's coherer tends to respond to all of the signals, or to give an unintel-
ligible mixed record. It is true that if the waves
from the different competing sending stations
have different lengths, that length of wave which
most nearly conforms to the quadruple of the
equivalent ship's antenna height will preponder-
ate in strength. Nevertheless, in an untuned
system, the other waves are likely to interfere.
This is partly because, as we have already seen,
a simple sending antenna, not tuned in connection with an auxiliary discharging condenser,
tends to emit very short wave-trains, that are
virtually but solitary waves with tails to them;
and partly because a simple coherer, at the base
of a simple receiving antenna, does not admit of
much resonant building up of voltage, even with
long wave-trains.

## *Auditory Selection*

This interference and jumbling together of signals from different sending stations in the neighborhood was soon found to constitute a menace to wireless signalling on any extensive scale, especially with the coherer type of receiving instrument. With other types of receiver which operate through a telephone, less trouble from interference is liable to be felt. This is for the reason that in a telephone receiver the signals usually have a buzzing sound, or possess a definite semi-musical tone. The pitch of the tone corresponds to the frequency of the induction coil vibrator at the sending station; or, as it is termed, the *group-frequency; i.e.*, the number of impulses per second delivered to the sending induction coil when the sender's key is held down (see Fig. 33), or to the number of groups of waves emitted per second. As a general rule, different sending stations do not employ just the same group frequency, or pitch of vibrator, so that the characteristic buzz or tone of the signals heard in the receiving telephone is different for different stations. When, therefore, a number of stations are sending messages simultaneously in the neighborhood, an untuned receiving telephone set will render them all audible at once. If they have all exactly the same tone, it would be impossible to make anything of the jumble of signals; but if, as usually happens, the tones are appreciably different, the conditions resemble the jumble of sounds maintained in a reception room, when many individuals are speaking close by at once. It is often possible, with a little effort, to focus attention on one particular succession of tones and mentally to read the signals they contain, to the exclusion of all the others.

### Need for Resonant Selection

There is, however, a limit to the possibilities of deciphering one tone of telephonic signals from among a crowd. If a powerful antenna near a civilized seashore is connected untuned to a sensitive telephone employing receiver, a regular babel of signals is frequently to be heard. Some of these are from shore stations nearby, others from distant shore stations, and yet others from ships at sea. As we look upon the surface of a large lake, or of the sea, we usually discern waves or ripple trains, which are crossing, superposing, or intersecting in endless variation. A calm, unruffled water surface is the exception. The same conditions now apply to the atmospheric ether in civilized districts, so far as concerns long electromagnetic waves.   The atmospheric ocean is rarely quiescent.

### *Adjustment to Resonance*

In order to bring about sharply selective signaling, it is necessary that the receiving antenna and apparatus connected therewith should be tuned to one definite wave-length, so as to respond to waves of that length exclusively, and also that the sending station desiring to communicate solely with that receiver should be tuned to emit as long wave-trains as possible, possessing this particular wave-length.   Theoretically, it should be possible to adjust an antenna into resonance to a given wave-length within any desired degree of precision, so that if the waves received were one per cent. too short, or too long, they would fail to actuate the receiver.   There is, however, a limit to precision in practical tuning for various reasons.   About five per cent. above or below is usually regarded as satisfactory. That is, a receiving antenna and apparatus can be arranged to respond to a given wave-length of arriving signals, and not ordinarily to respond to waves five per cent. shorter or longer.   This means that the receiver would not ordinarily recognize, or report in the telephone or recording apparatus, any passing waves outside of these limits.

There are various ways of connecting, sending and receiving antenna circuits in order to tune them. One such way is indicated in Fig. 62 for the sending apparatus, and in Fig. 63 for the

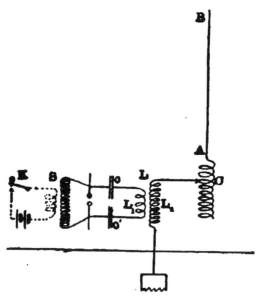

Fig. 62.—Particular Set of Tuned Sending Connections.

receiving apparatus. By means of a suitable switch, or group of switches, the change may be made from one set to the other, that is from sending to receiving, with one and the same antenna. Referring to Fig. 62, A B is the antenna or wire leading thereto, C an adjustable coil for virtually

altering the equivalent height and wave-length
of the antenna. A high-frequency induction coil
of relatively few turns without any iron core, is
indicated at L, the secondary winding L, being
connected to the antenna, and the primary L, to
the secondary terminals of the spark coil S,

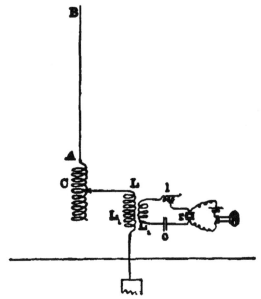

FIG. 63.—Particular Set of Tuned Receiving
Connections.

through adjustable condensers $c$ $c'$. The key K
is in the primary circuit of the spark coil S.

The circuit $s$, $c$, $L_1$, $c'$ is adjusted to oscillate
electrically at the required frequency and the
antenna path $B$, $A$, $C$, $L$, is also adjusted to

oscillate at the same frequency. It generally results that there are at least two different frequencies, and not merely one frequency of oscillation set up in such a system; but one of the frequencies is taken as the effective or working frequency, and the others are regarded as ineffective or merely parasitical.

Turning now to the receiving connections of Fig. 63, A B is the antenna wire and C an adjustable coil as before. L is a high-frequency induction coil. In the secondary circuit of this coil is an adjustable small coil l, an adjustable condenser c, and the receiver r, connected with the local telephone, or recording instrument. The antenna path is adjusted to oscillate at the required frequency, and the secondary circuit $L_s$, l, c, r is also adjusted to oscillate at this frequency Under these circumstances the whole receiving system is tuned to resonance with the single required frequency, or wave-length.

### Simultaneous Sending and Receiving with Aid of Differential Resonance

When tuning is carefully and effectively carried out, it may enable remarkable results to be accomplished. For instance, it has been found possible to receive messages over an antenna at, say, the foremast of a steamer, and at the same time to send messages in another wave-length, from an antenna at the mainmast, to a different and perhaps very distant station.   It is evident that this result would not ordinarily be possible with simple untuned apparatus at both antennas, nor would it be possible with tuned apparatus, if the frequency and wave-length of both the sending mast and the receiving mast were the same.

### Increase of Transmission Range by Means of Resonance

The advantages of tuning are found not only in the elimination of interference from extraneous signaling stations, but also in increase of sensibility and effective signaling range.   The tuning of the sending station connections permits of increasing the length of the train of waves following each discharge of the spark coil.   The sympathetic tuning of the receiving station connections permits of building up a resonant current strength in the receiver, due to the successive additions of impulses from the successive waves in the train.   By this means, signals which would be too faint to be recognizable if they depended upon the impulse of a single wave, become recognizable by the cumulative impulses of a number of successive waves.   There is good reason, therefore, for expecting that when a suitable high-frequency dynamo is developed for the continuous maintenance of power in the sending circuit and antenna, a further great increase in effective range of transmission will be rendered possible.

## Reduction of Atmospheric Disturbance by Means of Resonance

Another advantage derived from tuning is in the direction of minimizing the influence of atmospheric discharges. An antenna is a sort of lightning rod, and the taller and more extensive it is, the better atmospheric discharger it tends to become. The atmosphere contains electrically charged particles or free electric charges. Their presence may be accounted for in several ways that need not here be discussed. Consequently, an antenna is apt to receive a perpetual stream of little electric discharges from the layers of air near its top, to the ground at its base. This action is quite distinct from the much more powerful discharges which may occur over the antenna in the presence of a thunder-. storm in the vicinity. During such a thunder-storm it is usually necessary to stop signaling and to keep the antenna grounded. The little atmospheric discharges become objectionably noticeable in the receiver. and sometimes give rise to false signals. These continuous atmospheric disturbances are stated to be more noticeable and troublesome in the tropical than in the temperate zones, but they vary in intensity from day to day and hour to hour. On some occasions they are almost entirely absent, and on other occasions they are markedly prevalent. Distant thunderstorms and atmospheric discharges likewise produce noises in the receiving telephone, interfering more or less with received signals. Tuning of the antenna connections is capable of reducing atmospheric disturbances, although perhaps not of eliminating them entirely.

## Multiple Wireless Telegraphy by Means of Resonance

The results of tuning have even been carried further. It has been found possible to send two messages simultaneously over one and the same sending antenna, by connecting the antenna to two different coils, and auxiliary oscillating sending circuits, or to two different sections of one and the same coil. Again, it has been found possible to receive two messages simultaneously over the same receiving antenna by a corresponding connection to two oscillatory receiving circuits. This means that an antenna system may be arranged to emit two frequencies, or wave-lengths, simultaneously and independently. In the same way, one antenna may be arranged to resonate to each of two different frequencies or wave-lengths. By connecting a plurality of such oscillating circuits with an antenna, it is theoretically possible either to send or to receive an indefinite number of messages simultaneously, each in a definite appropriate wave-length. Up to the present time, however, but little use has been made of this possibility. A multiple system of wireless telegraphy is manifestly more complicated and difficult to maintain than a single system.

### Limitations of Communicability Through Resonance

Along with the advantages which pertain to a tuned or selective signaling system, there is one evident disadvantage, namely, loss of communicability. It is all very well for a ship which is tuned to receive waves say 300 meters (328 yards) long, or at a frequency of one million cycles per second, to be able to carry on communication with another ship, or a shore station, that uses the same wave-length; but a third station having a different tuning and producing waves, say 400 meters long, may desire to communicate with the ship, and not be able to do so, either from not knowing the particular wave-length to which the ship responds, or from not being able to alter his tuning thereto. It is partly for this reason that the ordinary wireless equipment on board ocean-going vessels is not sharply tuned. It is desirable that they should be able to speak to all comers within normal short range.

### German Practice

Under the auspices of the German government a dozen stations along the German coast are all tuned to emit waves of 365 meters (398.5 yards) in length, corresponding approximately to a frequency of 820,000 cycles per second. These stations have a signaling range of about 200 kilometers (125 miles) between each other, or about 120 kilometers (75 miles) from any one shore station to ships in its neighborhood. The ships are also tuned to this wave-length. Under these conditions the shore stations take precedence. When a shore station calls, ships within range are instructed not to speak unless called. Ships are also instructed not to call a shore station needlessly, or beyond the normal 120 kilometer (75 miles) range. By the observance of such regulations mutual advantage is subserved,

as well as the keeping of ethereal peace.   A wireless-telegraph etiquette is thus gradually becoming evolved.

### Difficulties in the Way of Using Reflectors or Lenses for Wireless Light

It would clearly be of great advantage if, instead of radiating waves from a sending station
to all the thirty-two points of the compass at
once, there were some convenient means of
channeling the waves into a beam, capable of
being sent into any desired direction.  This
would not only save wasted power, but would
also save needlessly stirring up the ether into
noisy signals in outlying regions where other
parties are trying to talk.   At first sight it would
seem that because we can readily accomplish this
result with short-length luminous waves, as in
the search-light beam for instance, we ought likewise to be able to accomplish it with the long
waves of wireless telegraphy.  The trouble is,
however, that optics leads to the general law that
both reflectors and lenses must be large with
respect to the wave-length of the waves they bend
into parallelism.  Thus, with half-micron waves
of light, a tiny mirror, no larger in diameter than
one millimeter (1-25th inch), would cover 2,000
wave-lengths.  If the mirror had a diameter of

less than half a micron ($\frac{1}{50000}$ inch) or less
than a wave-length, it would fail to serve properly
as a reflector.   In the same way, when we deal
with, say, 400-meter (437 yards) waves of invisible
polarized light, it would take a metallic surface of
more than 400 meters (437 yards) *square* to make
a serviceable reflector, and such sizes are pro-
hibitive.   A sending station situated in front of
a very steep overshadowing cliff, of fairly conducting surface, might have its shore side sheltered and its sea-going waves strengthened by
the reflecting surface of the cliff; but such locali-
ties are not always forthcoming;  nor perhaps is
the surface soil sufficiently conducting to provide
in most cases a satisfactory reflector.   It is well
known that a ship getting behind such a cliff may
be sheltered on that side from arriving signals,
or in other words that such cliffs may throw long
electromagnetic shadows beyond them.   On the
other hand, a vessel lying in a harbor enclosed by
hills which do not rise abruptly, but slope to the
harbor, will often receive wireless messages from
a direction over the hills, the waves in such cases
running down over the slope.

Although no marked degree of success has
hitherto attended the erection of reflecting verticals, or mirror surfaces, behind sending antennas, yet care has to be taken that a receiving

antenna is suspended reasonably clear of high conductors capable of casting a shadow. Thus a receiving antenna wire suspended behind a steamer's funnel of steel, or immediately behind a steel mast, would be likely to have its signals much weakened, if not entirely absorbed. For this reason an antenna is usually suspended from a rope between two masts, in such a manner as to hang free from both.

Some experimental progress toward the space direction of emitted waves has been recently announced by the addition of a horizontal antenna on the top of, and proceeding from, a vertical antenna. This construction is equivalent to taking say an 80-meter (87.5 yards) antenna, and instead of setting it entirely erect, placing 30 meters (32.8 yards) vertical and then carrying the remaining 50 meters (54.7 yards) out horizontally in a straight line overhead. In such a case the radiation is partly polarized in the horizontal plane, and partly in a vertical plane, the resultant being a sort of combination, or elliptical polarisation. Such waves are subject to different degrees of attenuation in different directions. It is found that the radiated waves are strongest in the direction opposite to that of the horizontal offset; so that this offset in the antenna should be made away from the direction

in which it is desired to send the waves.   In the
direction of the offset, the waves are weak and
they are weakest of all in a direction nearly at
right angles to the offset.

In time, it is to be hoped and expected that all
large vessels will carry wireless telegraph apparatus, and that all lighthouse stations along the
ocean shores will be equipped with invisible as
well as visible beams.   The invisible beams
would be perceptible by electromagnetic apparatus far out at sea, and also in foggy or hazy
weather.   This system of coast protection for
arriving vessels is already commencing.

### Desirability of Means for Determining Wave Directions at Sea

When a wirelessly equipped steamer ap-
proaches the coast of Europe or of America, she
is apt to be apprised of the proximity of land in
thick weather by the reception of signals from
coast wireless stations.   It is not easy to determine, however, the direction whence the signals
come, or the bearing of the station.   It would be
advantageous to be able to orient the received
signals, or to determine their direction of advance.
By suspending two or more antennas far apart,
near the bow and stern of the vessel respectively,
and bringing them into communication through

suitably tuned apparatus, it might be possible by swinging the ship to find the approximate bearing of the shore station. When the antennas were in line with the station, the signals would reach a maximum (or minimum), and when the line joining them was at right angles to the bearing of the station, the opposite condition should occur. Such a procedure is, however, objectionable, since it requires the ship to be stopped for observations, or at least turned erratically off her course. It seems likely that the first method to be found available will be for each shore station to erect a system of radiating wires, and to determine the direction of the ship by tests conducted upon alternate wires or pairs of these wires. The shore might then inform the vessel of her bearing, which would be the next best to the ship's determining the shore station bearing herself.

# CHAPTER XIII

## MEASUREMENTS OF ELECTROMAGNETIC WAVES

### *Importance of Determining Wave-lengths*

WE have already seen in Chapter VI that the length of the waves thrown off by a simple vertical rod oscillator, discharging into a perfectly conducting level ground, is four times the height of the rod; so that in the case of a simple unloaded antenna of the type represented in Fig. 26, we can form a close estimate of the wavelength emitted. But antennas are usually loaded by coils, condensers, or other apparatus, so that it becomes impossible to estimate their emitted wave-length with any degree of reliability. Moreover, the wave-length, which is not of great importance in simple untuned signaling, becomes of great practical importance in selective signaling by the aid of resonance. If we can readily measure the length of the waves thrown off by an antenna, we can proceed in a rational and straightforward way to tune the system, and also to produce any required wave-length within reach.

*Method for Determining Wave-lengths.*

The wave-length of an oscillating system is measured by bringing a portable, adjustable oscillation system into electromagnetic communi· cation with it, and adjusting the portable system until resonance is observed therein. At reso- nance, the wave-length of the tested system and of the portable system is the same. The wave- length of the portable system is determined, or computed, from its adjustments, and the wave- length of the tested system therefore becomes known.

The simplest kind of an adjustable oscillating system consists of a circuit containing a coil of wire and a condenser. Either the coil may be adjustable in its length and virtual number of turns, or the condenser may be adjustable in the extent of its effective surface, or both may be adjustable independently. The electromagnetic length of a coil is conveniently measured and stated in centimeters or meters. The *capacity* of a condenser is also conveniently stated in the same units of length. When the oscillating cir- cuit is adjusted to resonance, the length of its wave is the circumferential length of that circle whose radius is the geometric mean (square root of product) of the coil meters and condenser meters. Thus, if the circuit was in resonance when the coil was adjusted to 1,000 meters, and the condenser to 40 meters, the geometric mean of these two lengths would be 200 meters and a circle with this radius would have a circumference of 1,257 meters (1,375 yards) which is the wavelength of the resonant circuit.

### Means of Determining Resonant Condition in Portable Circuit

In order to recognize when the portable oscillation circuit has been adjusted to resonance, a *galvanometer, ammeter,* or current-indicator suitable for high-frequency to-and-fro currents, or oscillations is placed in the circuit. ·Fig. 64 shows one form of such an ammeter. It consists essentially of a short length of fine platinum

FIG. 64.—Hot-Wire Ammeter or Current Measurer.

wire, whose ends are secured to fixed supports. The center of the suspended wire is fastened to a thread that runs over a delicately supported pulley, carrying a pointer or index. If a rapidly oscillating electric current passes over the fine platinum wire, the wire is thereby heated and elongates. The elongation is detected by the sagging of the thread fastened to the center, and the degree of elongation, duly magnified, is caused to indicate the effective strength of the current. Care has to be taken that the current passing is not strong enough to overheat or melt the platinum wire. If such an instrument is inserted in a resonant oscillating circuit, the current indicated will become a maximum when resonance is produced.

### Thermo-galvanometer

A much more sensitive galvanometer or cur-·rent indicator depending also upon the heat pro-

FIG. 65.—Delicate Thermo-Galvanometer.

duced by current in a fine· wire, is seen in Fig. 65. A loop of fine platinum wire is supported close beneath, but not touching, the soldered junction of two wires of different metals, forming a circuit which is delicately suspended in a permanent magnetic field. When current passes through the fine platinum wire, some of the heat produced is communicated to the soldered junction. This raises the temperature of the junction and produces a steady *thermo-electric force* or voltage, as long as the elevation of temperature persists. The voltage sets up an electric current in the local circuit of the suspended loop, and the loop twists or tends to deflect sideways about the suspension, in a manner and to a degree which is greatly magnified by an attached mirror and a beam of light reflected therefrom. By means of this *thermo-galvanometer* very feeble oscillating currents may be measured. Some of the best measurements yet published concerning the strength of signals received at different distances from the sending station have been obtained by its use.

### Disk Galvanometer.

Another form of high-frequency galvanometer is indicated in Fig. 66. It consists of a coil of insulated wire having comparatively few turns, and supported at the center of the instrument, connected in the circuit under test. Inside this fixed coil is delicately suspended a little silver disk $s$, to which is attached a small mirror $m$, the two being carried on a fine quartz fiber about 30 cms. (12 inches) long. The suspension is illustrated separately in the figure. When a high-frequency current alternates in the fixed coil, a feeble alternating current of the same frequency is set up in the silver disk, and the electromagnetic attraction between the current in the coil and the current in the disk causes the disk to twist, or

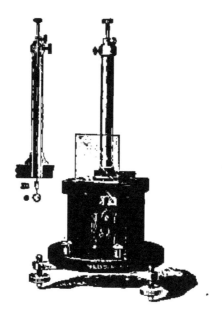

Fɪɢ. 66.—Oscillating-Current Galvanometer.

deflect about the axis of suspension. The deflection is magnified and rendered measurable by a beam of light reflected from the mirror *m*.

The portable oscillation circuit is brought into electromagnetic communication with the tested circuit, either by actual contact at two close points, or by means of a little induction coil of very few turns.

## Wave-measuring Helix

Another type of portable resonant circuit is made in the form of a helix, or close spiral, of fine wire wound upon an insulating rod or cylinder. Such a circuit has a greatly reduced or minified wave-length. That is, a wave which would occupy a length of say 50 meters (54.7 yards) of straight wire, would perhaps occupy only 15 centimeters (6 inches) of this wire wound in a curl or helix. When the antenna to be tested is in excitation, one end of this helix is presented near to but not touching the antenna, and the length of helix which will set up resonance is ascertained by running a grounded conductor along it. Resonance is in this case detected by the formation of a glow or brush discharge, either from the end of the helix itself, or within a small vacuum-tube placed adjacent thereto. A scale marked along the rod then enables the observer to read off the wave-length directly.

# CHAPTER XIV

*The Ocean, the Kingdom of Wireless Telegraphy*

WIRELESS telegraphy has already come into widely extended use over the ocean. It has not come as yet into extended use over land. The reasons for this are evident: Wire telegraphy has already held undisputed sway over land. Wherever there has been developed urgent demand for the telegraph, the wire has been run to meet it. But a moving ship cannot keep up wire communication with the land, except in the rare instances where a ship is employed in laying a submarine cable. Consequently, wireless telegraphy has absolutely undisputed sway over the surface of the ocean in reaching ships at a distance, or in reaching ships near by, when visual signals cannot be read, as at night, or in fog. Already, wireless telegraphy has done splendid work in maintaining communication with ships at sea. In a certain sense wireless telegraphy has removed the sea, because the sense of isolation in a vessel out of sight of land is almost entirely lost when messages are received on board through the ethereal medium of the air. In a psychological sense, distance has been destroyed. Moreover, since the sea is the conducting medium or broad conductor for guiding wireless telegrams, we may say that the sea has ceased to divide countries and now connects them.

### Prospects of Submarine Cable Telegraphy

It has been much debated whether wireless telegraphy would render submarine telegraph cables useless and cable property valueless. Up to the present time it has not done so, and there is no immediate prospect of its doing so. Taking, for example, the islands of Great Britain and Ireland, these are electrically connected with North America by thirteen submarine cables, and with South America by three more. Altogether there are about sixty wires connecting Great Britain with other countries. These cables are pouring messages into the islands at an average rate each of, say, fifteen words per minute, day and night continuously. It would be an enormous undertaking to replace these cables by wireless telegraphy, without any reference to all the other parts of the world served by submarine cables. Up to this date, wireless telegraphy has probably aided submarine cables by bringing telegraph messages to and from ships at sea, for transmission by cable, more than it has injured cable telegraphy by sending messages over the sea that would otherwise have gone underneath.   On the other hand, however, it must be remembered that wireless telegraphy is still very young, and that it has already made far more progress in its brief lifetime than did wire telegraphy in the corresponding period of its life, sixty to seventy years ago.   If wireless telegraphy continues to advance in the future at the rate it has maintained in the past, it may be that at some distant future time submarine cables will cease to be laid, and their work surrendered to their wireless rival.

There has been already at least one case where wireless telegraphy has supplanted a submarine cable, and that is between the U. S. army stations of Fort St. Michaels and Safety Harbor, Cape Nome, across Norton Sound on the coast of Alaska, a distance of about 177 kilometers (110 miles).   The cable having been repeatedly broken by ice on this ice-bound coast, the telegraph service has been carried on for about two years continuously by wireless telegraphy.

There are also several short sea distances between islands, or between islands and mainland, which have recently been covered by wireless telegraph equipments instead of by submarine cable. One of these is between Port Blair on the Andaman Islands and the mainland, Burma, a distance of 491.1 kilometers (305.2 miles) under the auspices of the Indian government telegraph department, the traffic averaging about ninety messages per month each way. On the other hand, a number of important long submarine cables have been laid since the introduction of wireless telegraphy.

### Peculiar Difficulties Incident to Wireless Telegraphy.

One of the difficulties that long-distance wireless telegraphy has had to deal with, which will probably always have to be expected, is disturbance from thunderstorms in the vicinity of a station. Such visitations are not rare during summer seasons, and usually call for a temporary suspension of traffic.

A curious lesser difficulty that long-distance wireless telegraphy has to meet is the effect of sunlight upon the atmosphere. Messages can be sent and signals received much further by night than by day. The effect of sunlight on the atmosphere is apparently to make the air foggy for the long waves of invisible wireless light. The nature of the action has not yet been clearly demonstrated; but it is supposed that the energy of sunlight, or short-wave light, in impinging upon the ocean of air, either disrupts many air-atoms and ionizes them, or else injects streams of ionized matter from the sun, thereby leaving floating electric charges in the air. The passage of electric waves through ionized air causes work to be done in displacing the electrons, or electric charges, and such energy is absorbed from the stock of energy in the waves. The waves therefore become enfeebled, or absorbed, in a manner suggesting the action of fog upon ordinary light. The degree of enfeeblement during daylight hours is not uniform, and varies from day to day in a most fluctuating and apparently erratic manner. This means that in order to carry on wireless telegraph service during the worst atmospheric daytime conditions, there must be a considerable reserve of power over and above that necessary during the night time under the best conditions.

It has been found that the atmospheric absorption of electromagnetic wave energy occurs more generally and more powerfully in the tropical zone than in the north temperate zone, presumably on account of the greater relative intensity of solar radiation upon the atmosphere in the tropics. It is also reported that the absorption depends in a marked degree upon the length of the electromagnetic waves and falls off very rapidly for lengths of wave exceeding 3 kilometers (1.86 miles); so that for wave-lengths of 3.75 kilometers (2.33 miles) and upwards, corresponding to frequencies of 80,000 cycles per second and under, the atmospheric absorption is comparatively small. There is still uncertainty as to the nature of the atmospheric conditions which produce absorption; but the great and sudden changes in the strength of transatlantic signals, which reveal themselves in a few minutes of time, suggest the presence of invisible masses of ionized air, cloudlike in form, which may hover in the body of the upper atmosphere, springing into existence under the influence of sunlight and disappearing when the ionizing influence of sunlight is removed. It is also thought that there may be a close connection between the degree of atmospheric absorption and the amount of magnetic variation of the compass needle, judging from a comparison of records for the daily variation of transatlantic electromagnetic wave mean absorption, and for the regularly tabulated daily variation of the magnetic needle.

The subject of atmospheric absorption of passing electromagnetic waves * is a promising field for future investigation bearing upon meteorology and terrestrial magnetics.

### Transoceanic Wireless Telegraphy

Regular transatlantic wireless telegraph transmission has been introduced between Clifden on the coast of Galway, Ireland, and Glace Bay on the shore of Cape Breton Island, Nova Scotia, an oversea distance of 1930 nautical miles (3570 kilometers or 2220 statute miles). Messages exchanged between these terminal antennas are transmitted by ordinary wire telegraphy to other places on each side of the Atlantic ocean. It is stated that about 5000 words a day are regularly transmitted across the ocean in this way.

### Extent of Use of Wireless Telegraphy on Vessels

A large number of naval vessels of different governments are now equipped with wireless

* See an important paper on "Wireless Telephony" by R. A. Fessenden, Proceedings of the American Institute of Electrical Engineers, July, 1908.

telegraph apparatus. Wireless telegraphy played a conspicuous part in the naval maneuvers of the Russo-Japanese war. By its means a blockade was sustained of Port Arthur for many months by a Japanese fleet at a safe anchorage a considerable distance away.

Wireless telegraphy equipment has been placed on board steamers of the following lines crossing the Atlantic ocean:

> The Anchor Line Co.
> The Cunard Steamship Co.
> The Norddeutscher Lloyd Co.
> The American Line Co.
> The Allan Line Co.
> The Atlantic Transport Co.
> The Compagnie Transatlantique.
> The Red Star Line.
> The White Star Line.
> The Hamburg-American Line.
> The Belgium S. S. Line.
> The Scandinavian American Line.
> The Navigazione Generale Italiana.
> The Austro-American Line.

About 116 vessels of this transatlantic fleet are so equipped. There are two powerful wireless transmitting stations on the shores of the North Atlantic Ocean, one at Cape

Cod, Mass., and the other at Poldhu in Corn-
wall. It is becoming customary for a vessel
with a long-distance equipment to maintain
communication across the Atlantic with one of
these stations until she establishes communication with the other; so that at no time is
she outside the range of communication.

A wirelessly equipped vessel leaving New
York is in communication with the station at
Sea Gate, N. Y. This is carried until it is
exchanged for communication with Babylon,
L. I. This is carried until Sagaponack, L. I.
Next Nantucket, Mass., is taken, then Sable
Island, and finally Cape Race. Communication for the exchange of messages is thus
maintained for about seventy hours after leaving port, each of these stations being in permanent wire communication with the rest of
the world.

It is stated that in the year prior to Jan. 31,
1906, these American stations sent and received, with ships, altogether 15,000 messages
comprising over 200,000 words.

### Number of Land Wireless Stations

According to a report of the U. S. Navy
Department there are this year (1908) about
468 land wireless stations in different parts of

the world, either erected or projected, namely :—

| | | |
|---|---|---|
| Belgium ...... 2 | Argentina .... 5 | Hong-Kong .. 1 |
| Denmark ..... 9 | Brazil ........11 | China ........ 9 |
| Germany .....12 | Canada ......31 | Hawaii ....... 7 |
| France .......18 | Chili ......... 3 | Japan ........15 |
| Great Britain..46 | Colombia ..... 1 | Dutch E. Ind.. 5 |
| Holland ...... 6 | Costa Rica.... 2 | Asiatic Russia. 1 |
| Spain ........19 | Mexico ....... 6 | Egypt ........ 4 |
| Portugal ..... 1 | Panama ...... 2 | Morocco ...... 5 |
| Gibraltar ..... 1 | United States.111 | Mozambique .. 2 |
| Italy .........22 | Trinidad ..... 1 | Tripoli ....... 1 |
| Malta ........ 1 | Porto Rico.... 2 | Canaries ...... 1 |
| Montenegro .. 1 | San Domingo.. 2 | Ecuador ...... 2 |
| Norway ...... 5 | Tongking ..... 2 | Formosa ...... 2 |
| Austria-Hung. 7 | Uruguay ..... 2 | Guam ........ 1 |
| Finland ...... 2 | Zanzibar ..... 2 | Korea ........ 5 |
| Switzerland ... 1 | Australia ..... 5 | Nicaragua .... 2 |
| Roumania .... 6 | Cuba .........10 | Peru ......... 5 |
| Russia .......15 | Tobago ....... 1 | Philippines ... 4 |
| Sweden ...... 4 | E. India....... 3 | Siam ......... 2 |
| Turkey ...... 5 | Burma ....... 1 | |

According to the same navy department list, there are at this date some 340 mercantile vessels equipped for wireless telegraphy, including the Atlantic liners already referred to, and carrying flags of the following countries:

| | |
|---|---|
| Belgium ......... 10 | Holland ......... 10 |
| Germany ......... 38 | Italy ............. 10 |
| France .......... 8 | United States.....141 |
| Great Britain..... 86 | Canada .......... 27 |
| | Japan ........... 10 |

· The total number of recorded ship and shore stations is thus about 800, exclusive of many warships of various nations.

Each station possesses a definite " code-name " or " call-letter," which is usually a group of two, or even three, letters, such as BA (Babylon, L.I). Stations are called by their code names and they sign their messages with them. This is even more necessary in wireless than in wire telegraphy; because the distant station may be far beyond visible range, and be otherwise unidentified.

It is customary for a steamer to pick up communication with another steamer at a distance of say 150 kilometers (79.8 nautical miles, or 93 statute miles) and to carry on communication until the distance between the ships is, say, 250 kilometers (133 nautical miles, or 155 statute miles). Occasionally, however, messages are exchanged between a ship and shore at much greater distances than these.

# CHAPTER XV

WIRELESS telephony depends upon the same principles as wireless telegraphy; but differs therefrom in details connected with the nature and requirements of the electric telephone transmitter and receiver. It is necessary, therefore, to have a sufficiently clear understanding of the nature and mode of operation of electric telephony with wires, in order to follow understandingly the modified nature and mode of operation of electric telephony without wires.

The word *telephony* is derived from two Greek words signifying the *far-off* transmission of *sound*.

## Nature of Sound

All material substances are capable of being compressed and dilated; i.e., of being altered in density, by the application of suitable forces to them; although the facility for being thus compressed and dilated varies enormously in different kinds of matter. For instance, gases, such as the air, readily admit of being compressed or dilated, as in the manipulation of a concertina; while, on the other hand, many liquids, such as water, require so much more force to compress or dilate them that, until the year 1762, it was supposed that water was incompressible.

When a body is subjected to a rapid rhythmic variation in its density, it is said to vibrate, as when the metal in a bell is forced into rapid small alternations of compression and dilation by a hammer blow; or when the framework of a building is set into slight vibration by the passing by of a rapidly moving railway train. Such rapid vibrations, communicating themselves to the ear, usually by way of the air and the external ear orifice, excite in our consciousness the sensation of *sound*. From this standpoint, sound is a particular mode of sensation excited by vibratory disturbances of density in neighboring material substances. But it is also customary to call a vibratory disturbance sound, when this disturbance is capable of exciting the sensation of sound. Consequently, sound may mean either the particular mode of physiological sensation received through the ear, or the physical disturbances in a medium, say air, such as might give rise to the sensation.

## Difference Between Musical Sound and Noise

When the vibratory disturbance, or sound, in a material medium is non-rhythmic and irregular in repetition, the sensation produced is called non-musical sound, or noise, as when coal is emptied from a cart into a cellar. The impacts of the many falling lumps of coal with the cellar floor, or with each other, set the coal and the surrounding air into powerful incoordinate vibration of a jangling character. When, however, the vibratory disturbance is rhythmically repeated, the sound sensation produced is more or less musical. If a gong is struck, its metallic mass is thrown into rapid vibrations which may be readily felt by the hand, and which are rhythmic in character. These rapid and rhythmic vibrations are communicated to the air surrounding the gong, and produce alternate compressions and dilatations in the air. Such local disturbances in the density of the air do not remain fixed in the space where they are produced; but move off in all directions at a definite speed.

When these moving air pulses impinge upon
the eardrum of a listener, they cause the
eardrum to vibrate, and communicate to the
listener the sensation of a musical sound.

### The Nature of Plane Waves of Sound in a Speaking Tube

A few examples of the spherical expansion
of sound-waves have been considered in
Chapter II. We may here examine the particular case of sound transmission within and
along a straight pipe, such as a speaking-tube.
In Fig. 67, A A', B B', represents a short
length of a speaking-tube containing air,
through which a simple single musical note
is being transmitted acoustically. The actual
process is quite invisible; but we may represent the density and pressure of the air in different parts of the tube by the relative proximity of transverse lines. Where the density
and pressure are above normal, as at C and C',
the lines are heavy and closely crowded.
Where, on the contrary, the density and pressure of the air are below the normal, the lines
are dotted and are also separated out. At
the particular instant selected, as represented
in the figure, the air at the points C C' is in
the state of maximum compression; while at

D D' the air is in the state of maximum dilatation. Between $N_1$ and $N_2$ the air is momentarily compressed; while between $N_2$ and
$N_3$ it is dilated or rarified, to a corresponding degree; but with ordinary intensities of
sound, the differences of pressure between the
compressed and dilated portions of the air are
remarkably small.

It is difficult to present clearly to the eye
the variations of density and crowding in the
air by means of the crowding together of
transverse lines as in the upper part of Fig.
67. A much more convenient diagram for
this purpose is given in the lower part of the
figure; where the straight line OO' represents normal or undisturbed air-density and
pressure; while deviations above this line stand
for compressions, and deviations below the
line for dilatations or rarefactions. Thus,
the wavy line d $n_1$ c $n_2$ d' $n_3$ c' indicates the
condition of pressure and density of all the
air in the length of pipe A B, at the instant
selected.

The sound waves $N_1$ C $N_2$ D $N_3$ C do not
stand still but move along the tube, from the
speaker to the listener, at a steady speed, which
in free air is about 335 meters per second
(1100 feet per second); but which in a nar-

row tube may be somewhat less, owing to
friction with the walls, say 315 meters per
second (1032 feet per second). The air in
the tube does not move bodily. When a
speaker blows through a speaking tube, the
air moves bodily along it, and may actuate a
whistle when escaping through the distant
end. But when he speaks into the tube, the

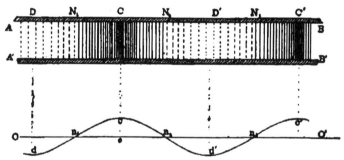

FIG. 67.— Diagram Indicating the Comparative Densities
of the Air in a Speaking Tube at a Particular Instant When Transmitting a Single Pure Musical
Tone.

sound of his voice may be carried through the
air without any bodily motion of that air.
Each air particle joins in the vibratory motion, and oscillates slightly to and fro in the
direction of the tube's length, about its mean
position of equilibrium. In other words,
when sound is transmitted, it is the disturbance in density which moves bodily along
in waves and not the parts of the substance in

which the disturbance exists.   The individual parts merely vibrate, being alternately closer together, and further apart, than in the quiescent state.

### Intensity or Loudness of Musical Tones

Musical tones, when steadily maintained one at a time, differ from each other in intensity, and in pitch.  The intensity or loudness of the sound sensation produced by a simple musical tone depends upon the amplitude of the vibration producing that sensation; that is, upon the maximum excursion of the air particles in their vibration from their mean position of equilibrium.   In Fig. 67 the amplitude would be measured by the distance $o\ c$. The loudness of a sound sensation increases with the amplitude of vibration; although not in simple proportion.  The sound of a steamer's whistle is often piercingly loud to a listener standing on the deck immediately in front of it; but becomes fainter with distance. This means that the amplitude of vibration of the particles of air is relatively great near the whistle; but becomes smaller as the distance from the whistle increases, and as the area of the sound-wave surface expands. The human ear is so sensitive to some sounds, that, according to accepted measurements, the sound of a whistle has been detected in air when the amplitude of disturbance at the listener's ear can only have been about 1 millimicron (1 m$\mu$ or $\frac{1}{25,000,000}$ inch.)

### Pitch of Single Musical Tones

The pitch of a single musical tone depends only on the number of complete vibrations, or cycles, of disturbance per second. A note of low pitch, like that of a deep bass voice, possesses relatively few vibrations per second; while a note of a high pitch, like that of a soprano voice in its upper register, possesses relatively many vibrations per second.

The number of complete to-and-fro vibrations, or cycles, of vibration per second, executed in a single musical tone is called its *frequency,* as in the case of electric vibrations referred to on page 63. A note of high pitch is, therefore, a high-frequency note, and a note of low pitch a low-frequency note. The human ear is able to hear sounds whose frequencies lie between about 16 cycles per second in the bass and about 16,000 cycles per second in the high treble; or over a range of some ten octaves, the limits of pitch audibility varying to some extent with different individuals.

The usual pianoforte keyboard is from $A_1$ of 27 cycles per second, in the bass, to $c^5$ of 4100 cycles per second, in the treble, or about $7\frac{1}{4}$ octaves. The ordinary range of pitch in the singing voice is somewhat less than 2 octaves, a man's baritone compass being commonly from A of 108 to f' of 316 cycles per second, and a woman's soprano compass from c' of 256 to a'' of 854 cycles per second. The fundamental tone of a man's speaking voice is usually in the neighborhood of 150 cycles per second with a wave length in air of about 2 meters (6.56 ft.), and that of a woman's voice near 300 cycles per second with a wave length of about 1 meter (3.28 ft.).

### Purity of Musical Tone

Contrary to what might be supposed at first thought, a pure musical tone in the sense of a single simple musical tone, cannot be produced by the human voice, is very difficult to produce artificially, and when produced, is not particularly pleasing to the ear. A musical tone produced by a trained voice is found to be not a single simple musical tone of the desired pitch; but a harmonious association of feebler higher pitch tones with the tone of desired pitch. The wave form of a simple musical tone is indicated in the line *d c d' c'* of Fig. 67. A close approximation to such a tone may be produced by mounting a tuning fork on a suitably shaped hollow chamber, or resonator. A flute may also be made to produce a fair approximation to a single

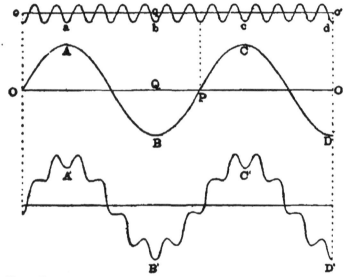

FIG. 68.— Composition of a Simple Musical Tone with an Overtone of Eight Times Its Frequency and One-Fifth of Its Amplitude Into a Resulting Composite Musical Tone.

musical tone. Ordinarily, what we describe as a single musical tone, is the association of that tone with a number of fainter tones of higher pitch; so that the wave shape is rendered complex. To take a simple example,

Fig. 68 shows at o A B C D the wave-form of a certain pure musical note, say middle c′ of the piano, with an amplitude Q B. Above this appears the wave-form of another pure musical note *a b c d* of eight times the frequency, and corresponding therefore to the triple octave, or c‴ above the treble clef. with an amplitude *q b*, one-fifth of Q B. If both these pure musical notes are sounded together, the resulting wave-form is shown at A′ B′ C′ D′. A trained ear listening to the composite note might detect both the *fundamental tone* of O A B C D and the fainter *overtone* or *harmonic* o a b c d, which would blend together harmoniously. Even an untrained ear might detect that the quality of the composite tone A′ B′ C′ D′ was different from that of the simple tone A B C D.

The wave shape of a composite musical tone may be altered in three ways :—

(1) By changing the number of associated overtones.

(2) By changing the relative amplitudes of associated overtones.

(3) By changing the relative positions, or "phases," of the overtones.

The quality of the tone as appreciated by the ear will be affected by changes (1) and

(2), but not by (3). In regard to change
(3), it may be observed that in Fig. 68, the
overtone has the negative, or downward,
amplitude $a$ at the moment when the fundamental tone has the positive, or upward, amplitude A; so that the composite tone wave is
diminished in amplitude at A' and C'; but
increased at B' and D.' If, however, the
ripple train $a$ $b$ $c$ $d$ were advanced though
half its wave-length, or *changed in phase* by
half a wave, with respect to the fundamental
wave O A B C D, the composite tone would
have the same quality to the ear, but its wave
form would have increased amplitude at A'
and C', with diminished amplitude at B'
and D'.

It usually happens that a source of musical
tones, such as a horn, trumpet or harp, produces, along with each fundamental tone, an
association of a number of fainter overtones,
whose frequencies are usually all simple mul-
tiples of the fundamental frequency. The
number and relative prominence of these
overtones give the distinguishing quality of
the note produced by each instrument. Thus,
a flute sounding middle c', produces relatively
few and feeble overtones. The fundamental
tone is heard almost pure. On the other

hand, the same note sounded on a violin would be accompanied by a large number of overtones, of successive frequencies 2, 3, 4, etc., times that of the fundamental note. As a general rule, the higher the frequency of an overtone, the smaller its amplitude; so that beyond a certain frequency the overtones tend to become inappreciable. Occasionally, however, particular overtones, such as the 7th — or 9th-frequency overtones, may be more prominent than their neighbors. The shape and physical conditions of the violin sounding-board tend to accentuate some overtones more than others. The reason, therefore, that the note, say " middle c'," sounds quite differently when sung by a voice, piano, or violin, lies mainly in the differences of associations of overtones, and in the corresponding wave shapes of the composite tones.

Moreover, the same sustained musical note sung by a trained singer, and by an untrained singer, may be very different, in spite of the fact that each may be producing essentially the same fundamental tone. In the untrained voice, there is likely to be a wavering, or unsteadiness, of pitch, or of amplitude, or of both, due to imperfect muscular control. There may also be an unmusical roughness, or noise, included in the tone, owing to the imperfect interaction of the vocal chords in the larynx. There is also likely to be an unpleasing association of overtones, both in regard to their relative amplitudes, and to their number; while the particular association of overtones may be varying, or wavering, from moment to moment in an unpleasant manner. In the note of the trained singer, we are likely to find, on the contrary, a sustained steadiness, either in uniformity, or in graded change, of loudness, an absence of roughness, or extraneous unmusical sound (noise), and also a pleasing association of evertones, brought about by the habitual formation of the vocal cavities so as to resonate with, and reënforce, harmonious components. Similarly, the difference in the quality of the same notes produced by a player successively, and with the same skill, on different violins or pianos, depends mainly upon their respective sounding-boards, and the resonating influence of these on the overtones. Some particular blendings of overtones in regard to number, or relative amplitudes, are more pleasing to the ear than others. The skill of the instrument maker is shown in the resonating qualities he is able to bestow upon the instrument when it leaves his hands.

## Tones in the Speaking Voice

We have already seen that the wave-forms of the sounds in the singing voice are complex in character, owing to the large number of different single tones that are ordinarily contained therein; but the sounds of speech are still more complex. In the sounds of speech we find *vowel-sounds* and *consonant-sounds*, as well as inflections and cadences of tone. The vowel-sounds are of a quasi-musical character, and the musical quality of a speaker's voice depends in large measure upon them. The inflections and cadences of speech are mainly variations in the fundamental tones of the vowel-sounds. The consonant-sounds are of different kinds, such as labial, dental, guttural sounds; but are mainly quick, sudden and explosive. The more prolonged vowel-sounds connect and are terminated by the more sudden consonant-sounds. The definiteness and intelligibility of speech resides principally in the consonant-sounds. Speech, deprived of its consonants, becomes a mere droning, or caricature of song.

Some of the consonant-sounds are feebler, or have smaller amplitude, than vowel-sounds. This is particularly the case with sibilants, such as *s, z, sz,* etc.   One of the most difficult words for a phonograph, gramophone, or telephone to repeat is " *specie.*"

Owing to the large range of frequency in consonant-sounds, and their frequent lack of amplitude, it is more difficult to reproduce articulate speech than vowel-sounds, or music. It may be possible for a phonograph, or telephone, to reproduce recognizable musical tones, when the reproduction of recognizable speech is impossible.

# CHAPTER XVI

## THE PRINCIPLES OF WIRE TELEPHONY

In the ordinary process of electric telephony by means of wires, the speaker talks in front of a "transmitter" such as that shown at T in Fig. 69. The essential elements forming this transmitter are indicated in Fig. 70. M is a hard rubber mouthpiece, usually provided with a perforated grid at the base, to prevent a pencil, knife or other pointed instrument from being pushed in, to the detriment of the delicately adjusted parts

**T**

FIG. 69.— Ordinary Desk Set of Telephone Receiver and Transmitter as Used in Wire Telephony.

beyond. An aluminum circular diaphram D is supported around its edge, and held in a soft rubber groove or gasket. At the center C, the diaphram is hollowed out to form a circular chamber. In this chamber are placed two carbon disks F and R, separated by granules of hard carbon. The front disk F is carried by the diaphram D. The rear disk R is fastened rigidly to a pin at the center of the solid metal back B. A thin mica disk A. is clamped between the diaphram and the rear disk, so as to close the chamber flexibly and maintain a moisture-tight seal. The front and rear disks are connected by wires to the terminals of the transmitter.

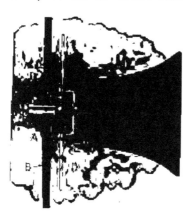

When the speaker's voice is directed towards the transmitter, the sound waves in the air enter the funnel or

FIG. 70.— Sectional View of Principal Parts of a Telephone Transmitter.

mouthpiece M, and impinge upon the aluminum diaphram D, which is set into vibration corresponding to the vocal vibration. The diaphram flexes and buckles to and fro very rapidly, as indicated diagrammatically by the dotted white lines in the figure. Since the solid metal back plate B is practically rigid, the rear carbon disk R stands fixed, and resists the vibratory force. Consequently, the particles of hard carbon C lying between the vibrating front carbon disk F and the stationary rear carbon disk R, are subjected to alternating compression and relaxation of pressure. These vibratory changes in pressure accompany the vibrations of the diaphram D, which as we have seen, follow the air vibrations of the waves of sound arriving from the speaker's lips. The powdered carbon F has the peculiar and valuable property that, when lying loose and uncompressed, it offers considerable resistance or obstruction to the passage of an electric current; whereas, when compressed and compacted, this obstruction is in considerable part removed. Consequently, a voltaic battery, connected to the transmitter, is able to send more current through the powdered carbon F each time that the diaphram D is moved inwards to compress the carbon, but is compelled to send less current each time that the diaphram D is moved outwards to release pressure on the carbon. Each vibratory motion of the

disk thus produces a corresponding vibratory
impulse of electric current in the wires
carrying the current from the battery to the
transmitter. It is as though the vibrating
disk governed a little throttle-valve, by which
electricity was alternately admitted to and cut
off from the circuit comprising the battery,
the wires, the transmitter, and any other instruments included therewith.

It is easily understood that the successive
vibratory movements of the diaphram are immediately followed by similar successive electric current impulses along the wires connected to the transmitter, owing to the correspondingly varying electric resistance of
the carbon particles F. The electric current
impulses move along the conducting wires as
invisible electromagnetic waves at very great
speed. If the diaphram D behaved perfectly,
it would faithfully repeat in its vibratory
movements each and all of the vibratory
movements of the impinging air particles. In
other words, a perfectly acting diaphram D
would follow all of the faintest ripples on the
back of the most complex wave-forms pertaining to the incident vocal sounds. Assuming such perfect behavior on the part of
the diaphram, the wave-forms of the vibra-

tory pressure communicated to the powdered carbon c would be the exact counterparts of the wave-forms of the vocal sounds uttered by the speaker. The effect of the corresponding changes in electric resistance in the carbon would be to produce electric currents whose wave-forms would all correspond with those of the vocal sound-waves. As a matter of fact, however, the diaphram D is never perfect in its behavior. It tends to develop favorite vibrations of its own, considered as a flat bell or gong, and it distorts more or less in its actual vibrations, the wave-forms of the air vibrations impinging on its surface. Nevertheless, the vibrations of the diaphram D follow those of the incident sound-waves sufficiently nearly for practical telephonic purposes, and the electric current waves, which closely correspond to the diaphram's vibrations, represent the vocal sound-waves fairly well.

By means of suitable delicate electromagnetic mechanism, the electric current waves in a wire telephone circuit can be made to photograph themselves, if care is taken to make them relatively powerful. With this object in view the circuit must be comparatively short: that is, it must not include many miles of wire, the instruments must be adjusted as delicately as possible, and the speaker must place his lips close to the mouthpiece of the transmitter and speak in a full clear tone. Many persons fail to make themselves clearly heard in ordinary wire telephonic conversation, because they talk into the circumambient air, instead of talking into the transmitter. A low tone of voice, with the lips nearly touching and fully opposite to the transmitter mouthpiece, is likely to be more effective in making the distant listener understand, especially in long-distance telephony, than loud shouting with the face directed away from, or to one side of, the transmitter.

Fig. 71 presents three " oscillograms " or photographs of the electric current waveforms in the articulation of the three syllables *cur, pea,* and *tea.** Except for the vibratory imperfections of the transmitter diaphram, above referred to, these electric wave pictures may be regarded as portraits of the soundwaves in the voice of the speaker that uttered those syllables. Beginning at A on the top line, the interval A B represents a small frac-

* From a paper on " Telephonic Transmission Measurements " by B. S. Cohen and G. M. Shepherd, Proceedings of the Institute of Electrical Engineers, London, May, 1907.

tion of one second of time, during which the
speaker uttered the syllable *cur*. First comes
the *c* consonant or sound of *k*, as a train of
about twenty small high-frequency waves of
very complex form. Then there is a brief
pause, during which the muscular adjustments
appear to be made for the following vowel-
sound *ur*, and finally we have about eight

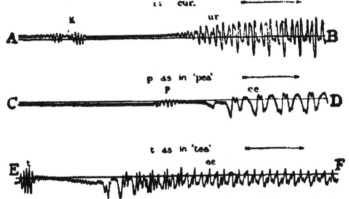

FIG. 71.— Photographs of Electric Current Waves in the
Transmission of Three Particular Vowel-Sounds.

complete fundamental waves of the vowel-
sound, judging by the recurring sharp peaks
below the line, with numerous associated
overtones that distort the fundamental wave
almost beyond recognition. We can imagine
that if the outline A B were accurately cut
into the surface of a wax phonograph cylinder,
the passage of the reproducing stylus over the

indented surface might cause the instrument to repeat this syllable *cur*.

Similarly with the syllable *pea,* as recorded along the line c D. First comes the consonant sound, then a brief pause, and then about eight waves of the fundamental vowel sound with a clearly visible prominent overtone ripple of perhaps four times the fundamental frequency. Again, at E F, in the syllable *tea,* there is first the brief explosive consonant, then a pause containing apparently a feeble high tone, or a group of high-frequency tones, and finally the vowel-sound which somewhat resembles the vowel-sound at C D.

## Changes of Wave-form in Telephonic Transmissions Over Long Wires.

When electromagnetic waves are delivered to a pair of conducting wires, in ordinary wire telephony, by the action of the speaker's voice on his transmitter, two changes occur in these waves as they are carried over the wires to the listener at the receiving end: namely (1) a diminution in the amplitude, or strength, of the waves, and (2) a different diminution in waves of different frequency. The first change is a mere weakening, like that of sounds heard at great distances in air. It is

called *attenuation*.  The second change means
that the different frequency components in
composite sounds are attenuated differently, so
that the shape of the current waves arriving at
the receiving end of the line is different from
that of the outgoing waves at the sending
end.   In general, tones suffer more attenuation the higher their frequency.   That is, the

FIG. 72.—Oscillograms of Singing Voice at Sending
and Receiving Ends of a Moderately Long Telephone Line.

FIG. 73.—Oscillograms of Singing Voice at Sending
and Receiving Ends of a Considerable Length of
Telephone Line.

fundamental tones are not weakened so much
as the over-tones.   The result is that the character of the transmitted sound is altered during the electric part of the transmission.

The relative influences of attenuation and
distortion in wire telephony are fairly well
presented in Figs. 72 and 73, which are taken

from the same paper as the last illustration.
In Fig. 72, A B is the oscillogram of the electric current waves at the sending end of a
telephone line, produced by a fairly high note
sung into the transmitter by a girl's voice.
The corresponding oscillogram *a b* beneath,
shows the electric current waves at the receiving end of the line. The line was not of
great length from a telephonic standpoint.
In the oscillogram A B, there are 19 fundamental waves, judging by the lower peaks.
These correspond to the frequency of the note
sung. There is also a prominent ripple of
three times the fundamental frequency, and
there are, besides, other overtones discernible
of yet higher frequency. In the oscillogram
*a b* from the receiving end, there are the same
number of waves, but the amplitude is reduced. That is, there is evidence of considerable attenuation. Moreover, there is evidence
of some distortion, because the outlines of the
received waves are not merely smaller than
at A B, but they are also smoother and
rounder, indicating that the ripples have been
attenuated more than the fundamental. This
is more clearly shown in Fig. 73, where C D
and *c d* are the oscillograms at the sending and
receiving ends of a fairly long telephone

circuit when the syllable *oo* was sung into the transmitter. Here 14 waves of fundamental frequency may be detected. The waves received at *c d* are not merely attenuated. If only attenuated, they would retain the exact shape of the waves of the sending end, on a smaller scale of amplitude. They have also been distorted. The sharp overtones and peaks in c d are absent in *c d*. The received waves have more of the fundamental and less of the overtones in their composition. They approach more nearly to the type of simple fundamental wave appearing in Fig. 67. To a listener on such a telephone circuit, the voice of the speaker might be intelligible; but would probably sound quite differently. It would be altered in character and would probably sound "drummy." This is a well known condition pertaining to wire-telephony over circuits that are electrically very long and distorting.

On arriving at the receiving end of the telephone circuit, the invisible waves of electric current are enabled to reproduce corresponding sound-waves by passing through the coils of fine insulated copper wire in a telephone receiver. One form of receiver suitable for wearing on the head, has already been described in connection with Figs. 50 and 51. A particular form of hand receiver is seen partly disassembled in Fig. 74. The hard rubber shell SS' receives the connecting wires at the narrow end, and clamps the thin ferrotype disk or diaphram D between its broad end S', and the hard rubber screw cover R. Inside the shell is held the magnetic system,

FIG. 74.—Internal Parts of a Telephone Receiver.

consisting of a pair of hard steel permanently magnetized bars *a b, c d* connected at *a c* by an iron yoke-piece and terminating at *b d* in a circular bridge-piece *g g* of non-magnetic metal. On the poles *b d* are mounted soft iron strips forming the cores of two small electromagnet coils, which are wound with many turns of fine silk-covered copper wire. When assembled, the diaphram D is clamped close to, but out of contact with, the soft iron pole-pieces of the electromagnets. These are kept magnetized under the influence of the permanent bar magnets *a b, c d;* so that the diaphram D is steadily attracted or pulled magnetically towards the soft iron poles when no current passes through the instrument. If now a current passes through the electromagnet coils in one direction, the magnetic pull of the permanent magnet is strengthened, thus tending to bend down or buckle the diaphram D, near its center. If, however, a current passes through the coils in the opposite direction, the magnetic pull of the permanent magnet is weakened, and the elasticity of the diaphram D tends to flatten the diaphram or diminish its bending down at the center. Each wave, or superposed ripple, of electric current sets up a corresponding up and down movement of the center of the diaphram, the edge of which is held fixed between the ring on S' and an opposing ring on the cover R. Rapidly succeeding current waves thus throw the diaphram into vibrations, which, if the system were perfect, would be identical with those of the transmitter diaphram at the sending end of the circuit. The electromagnetic vibration of the receiver diaphram D sets in vibration the air over the diaphram, and when the cover R is pressed against the listener's ear, the sound-waves are led through the air directly from the diaphram D to the eardrum. Unless the electric current waves are much stronger than are ordinarily employed in telephony, the amplitude of vibration of the diaphram D is so small as to be imperceptible except with the aid of very delicate instruments. It is generally less than 1 micron ( $\frac{1}{25,000}$ inch). Nevertheless, within this small range, the vibratory movement of the receiver diaphram corresponds to that of the transmitter diaphram under the influence of the speaker's voice, after allowance has been made for the electrical and mechanical imperfections of the system.

# CHAPTER XVII

## PRINCIPLES OF WIRELESS TELEPHONY

IN order that long-distance wireless telephony may be possible, by means of electromagnetic waves conducted over the earth's surface, it is necessary that an antenna at the sending station should radiate waves that are definitely related to the sound-waves emitted from the speaker's lips, and that an antenna at the receiving station should pick up these waves and utilize them in such a manner as to reproduce these sound-waves.

An ideally simple arrangement would be that the transmitter, actuated by the speaker's voice, should generate alternating electric impulses supplied directly to an antenna, that the antenna should radiate the energy of these impulses in electromagnetic waves, the waveforms of which would be identical with those of the actuating vocal sound-waves, and that the receiving mast-wire should be connected to ground through a receiving telephone, and operate the same by the electric current im-

239

pulses produced by the passage of the waves. Such an arrangement is, however, impracticable, because the power of the human voice is insufficient to generate electromagnetic waves capable of producing audible sounds at any considerable distance. Moreover, the frequencies that are serviceable in transmitting speech are relatively low, not necessarily exceeding 2000 cycles per second, and seldom exceeding 5000 cycles per second. An antenna does not radiate electromagnetic waves to any considerable extent until the frequency is raised to at least tens of thousands of cycles per second. Such frequencies extend beyond the limits of audibility.

The general plan that is adopted is to supply electric power to the sending antenna under such conditions as will permit of sustained radiation of electromagnetic waves. This power supply is modified in some manner by a transmitter, under the action of the speaker's voice. The electromagnetic waves radiate out, carrying with them the vocally imposed modifications, and the distant receiving antenna, in the path of these waves, is able to make their modifications audible as articulate sounds in the connected receiving telephone.

*Methods of Maintaining Continuous Radiation*

It has already been pointed out in Chapter IX that when a sending antenna is supplied with power from an induction coil, operated through a vibrator, the radiation· of electro-magnetic waves from the antenna is likely to be markedly discontinuous. For instance, if the vibrator delivers 200 electric impulses per second to the antenna, the latter may radiate a brief train of waves at each 200th part of a second, with relatively long intervening gaps of quiescence. It is manifest that such a type of radiation is ill adapted for wireless telephony, because during the utterance of any one syllable at the transmitter, there will be only a few groups of waves, emitted from the antenna in jets, with relatively long intervening pauses. In order to transmit articulate speech, the radiation from the antenna must be continuously sustained; or, if discontinuous, the discontinuities must be relatively brief.

Two methods have recently been developed for continuously sustaining the radiation from a sending antenna. The first method employs the electric arc. The second method employs a high-frequency, alternating-current, dynamo machine.

## The Singing-Arc Method of Setting Up Sustained Oscillation.

One variety of the arc-lamp method is represented in its simplest elements by Fig. 75. A dynamo D supplies direct current to an arc lamp A, through suitably adjusted electric resistances R R′, and "choking coils" C C′. The main function of the resistances is to steady and control the current supplied to the lamp; while the choking coils prevent rapid current oscillations from traversing the branch C R D R′ C′ of the system. Connected in parallel with the arc lamp is a branch cir-

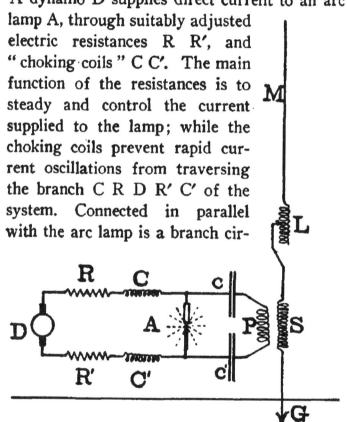

FIG. 75.—Arrangement for Maintaining Continuous Oscillation of an Antenna With the Aid of a Voltaic Arc.

cuit c P c', containing condensers c c' and a coil
P, which also forms the primary winding of an
induction coil, having its secondary S in the
sending antenna. The condenser capacity
and self-induction of the branch c P c' are
such as to favor the production of suitable
high-frequency oscillations. The mast wire M
is also tuned to the same frequency, with the
aid of the adjustable coil L, as described on
page 120, in connection with Fig. 37. The arc
lamp A serves to excite the oscillating-current
branch c P c' into sustained oscillatory action,
in a manner about to be described, and these
oscillations are imparted to the synchronously
tuned mast wire M, through the induction
coil; so that the mast wire is kept in continuous electric oscillation, and therefore, in
steadily sustained radiation of electromagnetic
waves. The energy carried off by these
waves is supplied by the dynamo D and by its
prime mover, say a gas-engine or steam-
engine.

The action of the arc lamp by which it ex-
cites oscillatory currents in the electrically
tuned branches c P c'. M L S G, is somewhat
complex in detail; but, in outline, is simple
enough. The solid cylinders, which support
the arc between their tips, offer comparatively

little resistance to the passage of electric current; but the vividly incandescent column of
metallic vapor, constituting the arc, offers a
considerable resistance, which depends in magnitude upon the strength of the current in the
arc. If the current is feeble, the arc is a
thin band of incandescent vapor, and offers
a relatively high resistance. If the current
through the arc is increased, the arc itself
swells and broadens, while, at the same time,
its resistance is lowered. In other words, a
stout arc, carrying a strong current, conducts
electrically better than a thin arc, carrying a
weak current.

If the arc lamp A is started with the oscillatory current branch c P c' removed or interrupted, a steady current will flow from the
dynamo D through the arc lamp and the coils
R C, R' C'. The arc will burn with a fairly
steady flame, as in an ordinary street arc
lamp. There will be no tendency to set up
high-frequency alternating currents in the system. When, however, the oscillatory branch
c P c' is applied to the arc, the condensers in
this branch take a sudden charge, or brief current impulse, which is deflected from the arc,
because the choking coils C C' do not permit
a sudden change of current to occur in the

supply circuit. The sudden diminution of current in the arc instantly causes the arc to shrink and rise in resistance; thus tending to throttle the current in the supply circuit D R C A C' R'. But the choking coils resist this sudden throttling of the current, and, in their endeavor to keep the current steady, they force electricity into the condensers, where it can go for the moment, after the conductive path through the arc is obstructed. The condensers thus become overcharged. Their electric elasticity speedily arrests the action, and forces electricity back through the arc, since the choking coils C C' resist all sudden changes. The current now builds up in the arc, and, as it does so, the arc column swells and conducts better. The arc resistance being thus reduced, the condensers over-discharge, being aided in this by the electromagnetic inertia of the induction coil primary winding P; while, at the same time, a strong oscillatory impulse is delivered from ground G to the mast wire M. The brief over-discharge of the condensers soon terminates, the current in the arc falls to the normal steady value, its resistance rises, and current is thereby deflected again into the condensers so as to charge them, at the same time inducing a

reverse impulse in the mast wire in synchronism with its natural period of swing. In this
manner a substantially steady flow of current
is delivered by the dynamo D, but alternately
in throbs or impulses to the arc A and to the
oscillatory branch c P c', the frequency of these
impulses being determined by the natural
period of the latter branch.

The amplitude of the current oscillations
that can be imparted in this way to the oscillatory current branch, and to the mast wire,
depends, other things being equal, upon the
change in resistance of the arc A with change
of current. If the arc changes greatly in
resistance for a given change in current, the
action above described will be powerful; while,
if the arc changes but little in resistance, the
action will manifestly be but weak. What is
needed, therefore, is an arc that is very sensitive in its resistance to changes of current.
This sensitiveness is found to depend partly
on the condition of the solid cylinders supporting the arc between their tips, and partly
on the condition of the gas in which the arc
is formed.

### Means Resorted to for Increasing the Sensitiveness of the Singing Arc

It has been found that the sensitiveness of the arc can be increased by substituting for the upper carbon rod a water-cooled metallic cylinder, and also by substituting for atmospheric air some other gaseous medium in which the arc is allowed to burn. Both hydrogen and illuminating gas have been employed. Moreover, it has been found advantageous to employ a plurality of sensitive arc lamps instead of a single arc lamp, in order to augment the action. In some cases, these arcs are all connected in series, while in others, they have been all connected in parallel.

### The Singing Arc

The sensitive type of voltaic arc flame above described is called the " singing arc," by reason of a curious and interesting inverse property which it possesses. We have seen that, when properly adjusted, the arc automatically charges and discharges the associated oscillatory branch c P c', by altering its volume and conducting power, in accordance with rapid variations of current strength. Such variations of breadth and volume in the

arc flame set up corresponding vibrations in
the surrounding air; so that the arc is able to
emit sounds. In the case of high-frequency
alternations, suitable for keeping an antenna
in electric oscillation, the sound would probably be inaudible, being above the limits of
audibility; but if the frequency is sufficiently
low, the arc can be made to give a fairly loud
tone. In fact, if a properly adjusted arc lamp
is supplied with a direct current, which has
passed through a suitably designed carbon
telephone transmitter, and musical sounds impinge upon the transmitter diaphram, the arc
will be able to reproduce them, even though
the arc may be at a great distance from the
transmitter. In such a case, the transmitter
produces rapid variations in the current sup-
plying the arc, in conformity with the incident
sound-waves. These current variations produce corresponding fluctuations in the volume
of incandescent vapor in the arc, which thereby exerts, in its turn, corresponding fluctuations in the pressure upon the surrounding air,
and so produces sounds in the same. An arc
lamp can thus be made to sing and reproduce
music. An arc so adjusted is called a singing arc. It is even possible to recognize vo
cal sounds reproduced by the arc, but the ar-

ticulation is seldom clear. In the case of adjusting an arc to reproduce sounds, the property of resistance variation in the arc vapor accompanying the sound is not brought into service; whereas in the application of the singing arc to exciting an antenna into oscillation, this property is of the first importance.

### Exciting Sustained Oscillation in an Antenna by Means of a High-frequency Alternator

The second method, mentioned above, for continuously sustaining the radiation from a sending antenna, employs a specially constructed high-frequency alternating-current dynamo, or *alternator*. The alternators which are used in America, for electric lighting and power transmission ordinarily generate either 60 cycles per second, or 25 cycles per second, the former frequency being suitable for street arc-lighting, and the latter for power transmission. The natural frequency of an unloaded antenna, 50 meters (164 feet) in height, is in the neighborhood of 1,500,000 cycles per second; so that this would be the proper frequency that an alternator should generate in order to be in simple synchronism, or in resonance, with the antenna. By loading the antenna, however, as described in

Chapter IX, it is possible to reduce the natural frequency of the antenna, or the frequency of its free oscillation and radiation. By this means, it has recently been found possible to bring the natural frequency down to a limit which specially constructed high-frequency alternators can attain.

Fig. 76.— High-Frequency Turbo-Alternator.

Fig. 76 illustrates a high-frequency turbo-alternator set. On the right hand side at T is a de Laval steam turbine which, with the aid of step-up gear, drives the shaft of the alternator A at a speed of about 16,000 revolutions per minute. The revolving element, or rotor, of the alternator, comprises a pair of steel disks, in the periphery of which 300 radial groves are cut. One of these disks is shown in Fig. 77. Between the grooves of two such revolving disks is mounted a thin stationary armature frame with 600 radial slots, and a coil in each pair of adjacent slots,

the coils being then connected in series. Each revolution of the disks, with their 300 polar teeth or projections, produces 300 cycles of alternating electromotive force in the armature; so that at the speed of 16,000 r. p. m. there will be

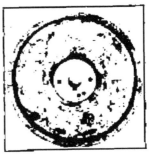

FIG. 77.—One of the Two Revolving Field Poles of the High-Frequency Alternator.

generated a frequency of 4,800,000 cycles per minute or 80,000 cycles per second. If now the armature is connected between the sending antenna and ground, and the antenna is tuned to this frequency, the alternator will be producing electric impulses in step with the natural electric oscillations of the antenna, and the system will be brought into full swing. The power developed electrically at the alternator terminals will be all radiated out from the antenna in electromagnetic waves, after deducting the heat losses which occur by the up and down movement of the electric current in the antenna. It has been found that the radiated power from a large and powerful sending antenna, when excited to resonance in the above manner, represents a load on the high-frequency alternator such as would be produced by a simple non-inductive resistance of 8 or 10 ohms. That is, the antenna, when in full oscillation, behaves as though it were grounded at the top of the mast through such a resistance.

### Receiving Circuit Connections

When, by either of the above methods, or by some other arrangement, a sending antenna is excited into steady radiative action, a corresponding steady emission of electromagnetic waves takes place at this antenna. Any receiving mast within effective range will then be able to pick up a steady electric disturbance in the antenna, caused by the continued passage of these waves as they run by. The disturbance will take the form of an alternat-- ing electromotive force, as described in Chapter VIII, and the frequency of the alternation will be identical with that of the sending antenna. The electrical effect of this alternating disturbance will be a maximum when the receiving antenna is tuned into resonance with that frequency, by suitably adjusting its load of inductance, or capacity, or both. It may be possible to use any type of sensitive and rapidly acting wave detector in the receiving antenna circuit connected with a receiving telephone. An electrolytic receiver, such as that described in connection with Fig. 46, is found to answer the purpose satisfactorily, and the receiving connections may be such as are indicated in Fig. 63. Under these conditions, a high-frequency received current will pass through the receiver $r$, and a current of the same frequency will also be set up in the coils of the receiving telephone connected to the receiver. The frequency of even a heavily loaded antenna is, however, far above the highest frequency that the ear can detect; so that nothing is heard in the receiving telephone, although a very considerable high-frequency alternating current may be maintained flowing up and down the receiving antenna through the receiver $r$. In order to break this silence, it is necessary to modify the oscillations of the sending antenna in accordance with the vocal sound-waves of the speaker, and to cause these modifications in the emitted

waves to manifest themselves in the receiving
telephone. Fig. 78 indicates diagrammatically a method of accomplishing this result.
The rapid oscillations O A B C D E F represent either the high-frequency alternating currents supplied steadily to the sending antenna

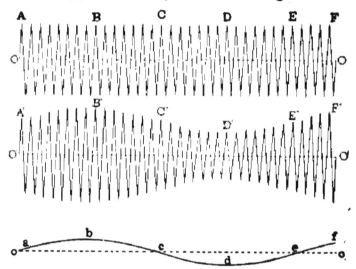

F_IG. 78.—Diagram Illustrating the Production of Audi-
ble-Sounds by the Modification in Amplitude of
Ultra-Audible Frequency Currents.

when no telephonic transmission occurs, or the
high-frequency waves which are steadily being
radiated, under that condition, from the sending antenna. If now the carbon transmitter
at the sending station can be made to alter
the amplitude of these outgoing waves,
in accordance with the diagram O′A′B′C′D′-

E'F', which performs periodic variations of thirty times lower frequency, then the telephone connected to the receiving antenna, as indicated in Fig. 63, may be regarded as giving vibrations of its diaphram corresponding to the variable amplitude high-frequency waves A'B'C'D'E'F'. This high-frequency may be beyond the limits of audibility; but the amplitude, wavering at the lower frequency, may produce an audible effect corresponding to the simple musical tone wave *a b c d e f.*

## Conditions Sufficient for the Transmission and Reproduction of Speech

In order, therefore, that articulate speech may be transmitted from the sending to the receiving antenna, and rendered capable of recognition by a listener at the latter, it is sufficient that an alternating current of ultra-audible frequency be steadily produced in the receiving antenna, and its apparatus, when the speaker is silent, and that when the speaker talks into the transmitter, the latter shall control the amplitude of the high-frequency waves, so that the complex wave-forms of the vocal tones may be developed in the shapes of the waves of amplitude. Under these conditions, if the degree of amplitude affected is sufficient, and the distance between the sending and receiving stations is not too great, the receiving telephone may reproduce the vocal tones of the speaker with sufficient power to make speech recognizable. In the case of Fig. 78, only a pure musical note could be expected to become audible, but if the transmitter had produced any association of tones within its compass, however complex, the same association might be expected to be bound up in the rapidly varying amplitudes of the successive outgoing waves and might be expected to be reproduced by the receiving telephone.

One method of controlling the amplitude of the high-frequency out-going waves, in accordance with lower frequency vocal sounds, employs a carbon transmitter in the main sending antenna path as indicated in Fig. 79. In this case, the antenna M having been adjusted into resonance with the high-frequency alternator A, behaves substantially like a non-inductive resistance to ground, that is, it

FIG. 79.—Transmitter in main Antenna Branch.

virtually closes the circuit of the alternator upon the radiation resistance of the antenna plus local connections. The carbon transmitter T, in the main circuit, alters the resistance of this circuit in conformity with the vocal sound-waves impinging upon its diaphram. The amplitudes of the high-frequency alternating current, and of the electromagnetic waves emitted from the

antenna, are thus caused to follow the wave forms of the speaker's vocal tones. It is not, however, necessary that the transmitter should be inserted in the main circuit. The transmitter may be placed in a circuit which is inductively connected with the main circuit, through the medium of an induction-coil such as is shown at I in Fig. 80. A condenser has also been used in the local transmitter circuit instead of the voltaic battery. In such cases, the variation of resistance in the transmitter circuit when the transmitter diaphram is disturbed by sound-

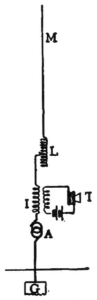

FIG. 80.—Transmitter in Inductively Connected Circuit.

waves is reflected inductively into the main antenna path, and serves to control the amplitudes of the emitted radiant waves. Another plan has been to place the high-frequency alternator in the primary circuit of an induction coil, the secondary circuit of which is connected with the antenna in the manner indicated in Fig. 36, care being taken to bring both the local oscillation branch A c B, and the antenna, into syntony with the high-frequency alternator, and with each other. This causes a steady stream of radiated waves of the same frequency to be thrown out from the sending mast and the amplitudes of these radiant waves is modified in conformity with the vocal tones of the speaker, by means of a transmitter, connected either in the antenna branch, or in some branch inductively associated with it.

The distance to which wireless telephony can be practically carried depends upon the amount of electric power that can be controlled in amplitude of current by the transmitter at the sending antenna and upon the limiting minimum of electric power which can be picked up from the radiated waves by the distant receiving antenna and utilized to operate the listener's telephone. The amount of power which the ordinary carbon transmitter controls in wire telephony may be less than one watt (equivalent to the power expended in lifting about ¾ lb. 1 foot high per second); whereas, in wireless telephony, much larger powers must be developed and controlled by the transmitter, on account of the scattering of this power in all directions from the sending antenna. The power that the transmitter is required to handle in long-distance wireless telephony may be hundreds or even thousands of watts. The ordinary carbon transmitter of wire telephony could not carry such an amount of power without becoming overheated and deranged. A special form of carbon transmitter designed to carry and control alternating currents up to 15 amperes is shown in Fig. 81. The mouthpiece M leads the incident sound-waves to a metallic diaphram which carries a short metallic rod fastened at its center. The rod passes through a hole in a metallic terminal plate and terminates in a platinum-iridium spade. Vibrations of the diaphram cause the spade to vibrate in a chamber packed with carbon particles, and having for walls two metallic terminal plates, which are separated by the white insulating ring. The metallic terminal plates are connected to the terminals T T and the spade to the terminal t. Water, admitted by the openings W W, is circulated through the two terminal plates to keep them from being over-

Fig. 81.— A Form of Water-Cooled Carbon Transmitter Employed in Wireless Telephony.

heated by the powerful alternating currents passing through the carbon.

### Freedom of Wireless Telephony from Distortion

It seems evident, both from the principles and the practice of wireless telephony, that although the radiated impulses which carry the vocal tones are subject to marked attenuation, being weakened, not only by the absorption of the waves into the surface of the imperfectly conducting land and sea, but also by the expansion of the waves into ever-increasing areas; yet they are not subject to the distortion which accompanies their transmission in wire telephony, particularly when the wires are placed close together in an underground cable. In other words, the sound-waves of wireless telephony get fainter as the range of transmission is increased up to the limits at present existing; but all of the tones transmitted become fainter in the same proportion; so that there is no indistinctness produced by the alteration of tone quality.

## Range of Wireless Telephone Transmission

The greatest distance reported at present for the transmission of recognizable wireless telephony in America is from Brant Rock, Mass., to Washington, D. C., a distance of 657 kilometers (408 miles). In Europe, the greatest reported distance has been from Monte Mario at Rome, Italy, and a vessel off the coast of Sicily near Trapani, an over-sea distance of over 500 kilometers (300 miles). Wireless telegraphy is still young, but wireless telephony

is younger still; so that the limits of range
to which the human voice can be carried, on
electromagnetic waves, are by no means yet
set. It would seem that the limits lie with
the amount of current and power which can
be handled by the transmitter, assuming that
no further improvements are made in direct-
ing the outgoing beam of electromagnetic radiation, in the antennas, or in the delicacy
of the receiving instruments. In one sense,
the extension of the present range from a
few hundred kilometers to the antipodes, or
half way around the world (20,000 kilometers, or 12,000 miles), would be less wonderful than the already accomplished feat of
reproducing recognizable speech at the range
now attained; because the extension of the
range of speech to the antipodes is a matter
of degree; whereas the achievement of wireless telephony to a range of even 100 kilometers (60 miles), is a wonderful acquisition
in kind.

It is only reasonable to expect, however,
that the range of possible wireless telephony
will be less than, and gradually increase towards, the range of possible wireless telegraphy, because, in wireless telegraphy, the problem of communication is limited to producing

any recognizable type of signal that can be repeated in successive periods of dots and dashes, whereas, in wireless telephony, the problem of communication involves the more complex condition of reproducing, at the receiving antenna, waves that have been successively modified in a long succession, substantially in accordance with the sound-waves of a speaker's voice, either at the sending antenna, or at a station connected electrically with the sending antenna.

### Selectivity of Wireless Telephony

Just as it is possible to select at a wireless receiving telegraph station one particular series of waves emitted from a particular sending station, to the exclusion of other sending stations, by some method of tuning; so it is possible to select at a wireless receiving telephone station, one particular series of waves emitted from a particular sending station, to the exclusion of other sending stations. For instance, if A, B, C and D, are four wireless telephone stations, so located as all to lie within each other's range of influence, and if A desires to speak with B exclusively; while C desires to speak with D exclusively, it will suffice for A and B to communicate in the fre-.

quency of say 80,000 cycles per second (wave-
length 3.75 kilometers), and for C and D to
communicate in the frequency of say 90,000
cycles per second (wave-length 3 1-3 kilometers). That is, not only the generating
source (arc lamps or alternator) at A would
be tuned to 80,000 ∽; but the sending antenna system of A, and the receiving antenna
system of B, would also require to be tuned
to this frequency. When suitably tuned in
this manner, waves of frequency 90,000 ∽.
would fail to be detected by B's receiver, and,
therefore, all variations in the amplitudes of
such waves, capable of reproducing speech,
would be cut off from B's telephone. The
telephone at B would only hear the speaker A,
to which it was adjusted in syntony. With
sharply-tuned antenna systems, it would be
possible for a number of such telephonic con-
versations to be carried on selectively, each
employing a powerful series of independently
acting electromagnetic waves.

### Simultaneous Speaking and Listening

With the arrangements above described, it
would be necessary to employ a switch to
change the antenna connections of a wireless
telephone station from sending to receiving,

*i.e.*, from speaking to listening, alternately. With ordinary wire telephony, such a switch was required in a very early stage of the art; but no switch is at present needed for this purpose, the transmitter and the telephone be-

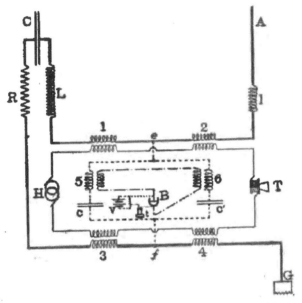

Fig. 82.—Connections for Simultaneous Speaking and Listening. Duplex Telephony.

ing always in the circuit simultaneously during conversation; so that it is possible both to speak and to listen at the same time. An arrangement of connections has been devised for effecting the same result for wireless telephony, and is indicated in Fig. 82. The antenna A, with its tuning coil 1, is permanently

connected to ground G, through the secondary
windings of four induction coils, 1, 2, 3, 4 and
an *artificial antenna* L C R, consisting of a suitably adjusted combination of inductance, capacity and resistance. The high-frequency
alternator H is connected to the transmitter
T, through the primary windings of the four
induction coils. Under these conditions, if no
sounds are delivered to the transmitter, a
steady high-frequency alternating current is
sent through the four induction coils to both
antennas. The real and artificial mast wires
are thus both thrown into full electric oscillation. If the artificial antenna is properly
adjusted so as to balance the real antenna,
the four induction coils will mutually neutralize each other's influences, and no current
will flow through the dotted system 5, c, 6, c',
which connects the points e, f, and which connects with the receiving telephone t. Again,
if the speaker talks into the transmitter T,
the amplitudes of the high-frequency alternating currents will be varied in accordance
with his vocal tones, but the power will be
equally divided between the real antenna A
and the artificial antenna L C R. The power
in the real antenna will be expended in radiated electromagnetic waves, after deducting in-

cidental losses in heating the mast wires. The power in the artificial antenna will be expended in heating the resistance R, which represents a radiation resistance, after deducting incidental losses in heating L and C. The real antenna may attain a height of 100 meters (328 feet), or more and may cover a considerable area of ground surface. The artificial antenna is a small affair that may be put inside a cupboard of one cubic meter space (26.4 cubic feet).

If electromagnetic waves are received at the real antenna A, of the same frequency as that to which it is tuned, they will develop an alternating current of that frequency passing to ground from the antenna through the points e f, and the dotted receiving system between them. Some of the current will pass also through the artificial antenna L C R; but will do no harm except in weakening the effect on the receiver. The divided primary receiving system 5c, 6c', is called an *interference preventer*. All of these four elements are independently adjustable. The secondary receiving system connects the two secondary coils through the liquid barretter or wave detector B, which is also connected to the voltaic battery v through the receiving telephone t. By suitably differentiating the two oscillating-current receiving branches, it is possible to tune the secondary system to respond loudly to the selected frequency, and to the practical exclusion of all others. The result is that the receiving telephone t is prevented from receiving any part of the locally generated high-frequency currents passing through T, owing to the differential balance between the four coils 1 2 3 4 and between the two antennas A and L C R. It will be silent to those currents, whether the transmitter T is spoken to or not; but the receiving telephone t is able to receive the incoming electromagnetic wave disturbances reaching the real antenna, because these disturbances are not destroyed by the differential balance. In this manner, both speaking and listening may continue simultaneously, as in ordinary wire telephony, although there is some weakening of the received currents, and also half the power available for sending out waves is absorbed locally as heat in the artificial antenna. That is, more power must be used with the arrangement of Fig. 82, for the same limiting range of recognizable telephonic communication, than with alternate speaking and listening.

The system of connections indicated in Fig.

82 is likewise available for duplex wireless telegraphy; that is, for the simultaneous sending and receiving of messages at one and the same antenna. It has been found that the artificial antenna, once adjusted to balance the real antenna, requires less change from day to day than does the "artificial line" employed in duplexing an ordinary wire tele-

FIG. 83.—Adjustable Condenser and Induction Coil Forming Elements of Interference Preventer.

graph line. This is apparently due to the fact that changes of weather have more influence in changing the electric conditions of a line hundreds of kilometers in length, than in changing the electric conditions of an antenna.

An adjustable air-condenser and an adjustable induction-coil for use in an interference preventer is shown in Fig. 83.

## Relaying Telephonic Currents to and From an Antenna

In telephoning wirelessly from a ship to a ship, or between a wireless shore station and a ship, the persons conversing together are

FIG. 84.— Automatic Telephone Relay.

close to their respective antennas; but when one of the persons is on shore at some place telephonically connected with, but remote from, the wireless telephone station on the sea coast, and wishes to speak to a person on a ship within range, it is necessary for his conversation either to be repeated by the operator at the coast station acting as intermediary; or to be repeated automatically to and fro by relays at the coast station. Fig. 84 shows a relay designed for this duty. It consists essentially of a telephone receiver in which a little vibratory tongue is substituted for the usual vibratory diaphram. The tongue dips into a trough containing carbon particles with water-cooled walls, arranged substantially as in the transmitter of Fig. 81. The apparatus is in fact a telephone receiver directly operating a carbon transmitter. The receiver is connected at the coast station to the incoming telephone line wire circuit, and transmits directly to the antenna. Another relay receives from the antenna, and transmits back to the wire telephone circuit.

# INDEX OF SUBJECTS

www.ingramcontent.com/pod-product-compliance
Lightning Source LLC
LaVergne TN
LVHW011941060326
832903LV00045B/115